Semantic Breakthrough in Drug Discovery

Synthesis Lectures on the Semantic Web: Theory and Technology

Editors
Ying Ding, *Indiana University*
Paul Groth, *VU University Amsterdam*

Founding Editor Emeritus
James Hendler, *Rensselaer Polytechnic Institute*

Synthesis Lectures on the Semantic Web: Theory and Application is edited by Ying Ding of Indiana University and Paul Groth of VU University Amsterdam. Whether you call it the Semantic Web, Linked Data, or Web 3.0, a new generation of Web technologies is offering major advances in the evolution of the World Wide Web. As the first generation of this technology transitions out of the laboratory, new research is exploring how the growing Web of Data will change our world. While topics such as ontology-building and logics remain vital, new areas such as the use of semantics in Web search, the linking and use of open data on the Web, and future applications that will be supported by these technologies are becoming important research areas in their own right. Whether they be scientists, engineers or practitioners, Web users increasingly need to understand not just the new technologies of the Semantic Web, but to understand the principles by which those technologies work, and the best practices for assembling systems that integrate the different languages, resources, and functionalities that will be important in keeping the Web the rapidly expanding, and constantly changing, information space that has changed our lives.
Topics to be included:

- Semantic Web Principles from linked-data to ontology design

- Key Semantic Web technologies and algorithms

- Semantic Search and language technologies

- The Emerging "Web of Data" and its use in industry, government and university applications

- Trust, Social networking and collaboration technologies for the Semantic Web

- The economics of Semantic Web application adoption and use

- Publishing and Science on the Semantic Web

- Semantic Web in health care and life sciences

Semantic Breakthrough in Drug Discovery

Bin Chen, Huijun Wang, Ying Ding, and David Wild

ISBN: 978-3-031-79455-1 paperback
ISBN: 978-3-031-79456-8 ebook

DOI 10.1007/978-3-031-79456-8

A Publication in Springer series
SYNTHESIS LECTURES ON THE SEMANTIC WEB: THEORY AND TECHNOLOGY

Lecture #9
Series Editors: Ying Ding, *Indiana University*
                Paul Groth, *VU University Amsterdam*
Founding Editor Emeritus: James Hendler, *Rensselaer Polytechnic Institute*
Series ISSN
Print 2160-4711   Electronic 2160-472X

Semantic Breakthrough in Drug Discovery

Bin Chen
Stanford University

Huijun Wang
Merck

Ying Ding
Indiana University

David Wild
Indiana University

SYNTHESIS LECTURES ON THE SEMANTIC WEB: THEORY AND TECHNOLOGY #9

ABSTRACT

The current drug development paradigm—sometimes expressed as, "One disease, one target, one drug"—is under question, as relatively few drugs have reached the market in the last two decades. Meanwhile, the research focus of drug discovery is being placed on the study of drug action on biological systems as a whole, rather than on individual components of such systems. The vast amount of biological information about genes and proteins and their modulation by small molecules is pushing drug discovery to its next critical steps, involving the integration of chemical knowledge with these biological databases. Systematic integration of these heterogeneous datasets and the provision of algorithms to mine the integrated datasets would enable investigation of the complex mechanisms of drug action; however, traditional approaches face challenges in the representation and integration of multi-scale datasets, and in the discovery of underlying knowledge in the integrated datasets. The Semantic Web, envisioned to enable machines to understand and respond to complex human requests and to retrieve relevant, yet distributed, data, has the potential to trigger system-level chemical-biological innovations. Chem2Bio2RDF is presented as an example of utilizing Semantic Web technologies to enable intelligent analyses for drug discovery.

KEYWORDS

drug discovery, semantic data integration, semantic analytics, semantic graph mining, semantic prediction

Contents

CHAPTER 1

Introduction

1.1   BACKGROUND

During the last two decades, drug candidate approval declined, and a number of marketed drugs were withdrawn due to severe side effects [Kola and Landis, 2004, Pammolli et al., 2011]. The low success rate challenges the current dominant drug discovery paradigm, which starts with the identification of a single target accounting for a disease, followed by the in vitro identification of selective compounds against that target with the expectation that the resulting optimized compound would exert desired therapeutic effects in vivo with tolerable side effects [Hopkins, 2008]. However, biological systems are complex; studies have shown that drugs tend to modulate multiple targets, affecting multiple pathways to treat a disease [Barabási et al., 2011]. In addition, drug response among patients is different due to genetic and environmental differences [Evans and Johnson, 2001]. The fact that the molecular mechanisms behind many marketed drugs are poorly understood is making the current drug discovery approach extremely challenging.

Therefore, research is moving towards the study of drug action on biological systems as a whole. In 2007, Oprea et al. proposed developing cheminformatics tools that could easily be integrated with bioinformatics tools, with the goal of creating Systems Chemical Biology. In 2008, Nature Chemical Biology published a special issue on chemical systems biology, in an effort to bring chemical biology and systems biology together [Editorial, 2008]. Starting in 2008, NIH began holding Quantitative and Systems Pharmacology (QSP) Workshops to gather the leaders in academia, industry, and government to discuss the status of the discovery, development, and clinical use of therapeutic drugs [Sorger et al., 2011]. Many pharmaceutical companies started exploring systems-based drug discovery pipelines, for instance, phenotypic screening [Young et al., 2008] and pathway-centric drug discovery [Fishman and Porter, 2005]. Since this is an emerging area, the definition, or even the term, has not been agreed upon [Editorial, 2008, Oprea et al., 2007, Sorger et al., 2011].

Nevertheless, essentially Systems Chemical Biology means, "integrative analysis of the interactions between drugs and biological systems with the aim of understanding the behavior of the system upon drug treatment" (see Figure 1.1). From the perspectives of informatics, Systems Chemical Biology includes two broad research topics: (1) the representation and integration of multi-scale datasets that make integrative analysis possible; and (2) the discovery of new knowledge from such integrated datasets.

Recent advances in chemical and biological sciences have led to an explosion of new public and private data sources regarding diseases, genes, genetic variations, proteins, chemical compounds, drugs, and their associations, providing an unprecedented opportunity to conduct integrative studies. For example, PubChem BioAssay (`http://pubchem.ncbi.nlm.nih.gov/`) and GEO (`http://www.ncbi.nlm.nih.gov/geo/`) curate over 59 million activity data points and 650,000 samples, respectively. Other than structured data, relevant unstructured data sources (including research literature, patent reports, news, etc.) are rapidly increasing. However, these data are isolated; for instance, a compound is tested against one protein in an assay, or the same compound is analyzed in a microarray for gene expression changes; two experiments are deposited in two separate databases. Logically, they ought to be linked in an integrated knowledge graph, to which we might apply intelligent data mining approaches, to gain insights into the complex functions of the examined biological systems and the actions of chemical compounds or drugs on them.

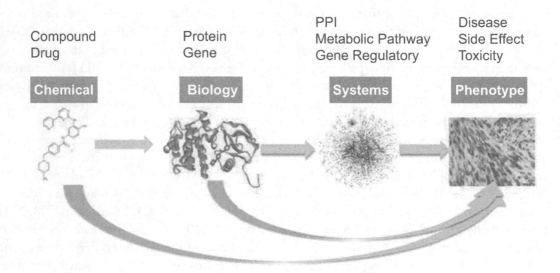

Figure 1.1: Schema of Systems Chemical Biology. Chemical compounds interact with proteins or genes that are functioning in biological systems. The modulation of their activities may affect the function of the systems, resulting in a change of phenotype. Datasets involved include (but are not limited to) chemical compounds, protein targets, genes, metabolic pathways, regulatory networks, diseases, side effects, and related factors.

Building an integrated knowledge graph involves the creation of large semantic networks that link multiple heterogeneous sources. It is a challenging task because of the following significant barriers.

1. Data formats are heterogeneous (e.g., text file, XML, relational database, CSV, HTML, etc.), and are structured or unstructured. A universal format is required.

2. Data types are heterogeneous (e.g., drug, compound, protein, gene, disease, etc.), and the size of each type of data is expanding quickly, making it impractical to build a centralized repository.

3. Many of the needed data sources cover similar data but from different perspectives. Semantic annotation and alignment is desired to merge homogeneous entities.

4. Contradictory facts and false relations are common; an appropriate approach is needed to manage them and track their provenance.

5. A new data model is needed to represent semantics of data and connections between data.

The Semantic Web [Berners-Lee et al., 2001] was envisioned to enable machines to process the Web of distributed data. It aims to build a framework allowing easy data sharing across multiple domains or different applications via open standards over the World Wide Web. As such, it has considerable potential to overcome the above mentioned barriers and facilitate Systems Chemical Biology studies. A number of Semantic Web standards have been developed including: the Resource Description Framework (RDF) (`http://www.w3.org/RDF/`), a universal data format used to represent and link data as graphs; OWL (`http://www.w3.org/TR/owl-features/`), an ontology language for defining concepts and relations; and SPARQL (`http://www.w3.org/TR/rdf-sparql-query/`), a query language and a protocol used to query the decentralized Web of data. Recently, efforts have been made to publish unstructured data in semantic formats, develop ontologies for annotating data, and create algorithms for discovering new knowledge from linked semantic data. Overall, the Semantic Web has demonstrated its utility in life sciences, healthcare, and drug discovery [Neumann, 2005, Wild et al., 2012].

The vision of the Semantic Web is to connect data with machine-processable semantics using World Wide Web protocols or standards. In the end, a giant "knowledge graph" of linked data on the Web can be formed. The data are not necessarily openly accessible, although it is encouraged; proprietary data can be included as well. The Semantic Web provides technologies to represent and manage data on the Web in a distributed and flexible way so that data can be connected and innovative algorithms can be applied to explore connected data and discover new insights.

In this chapter, we survey the state of the art of the Semantic Web and discuss the challenges of introducing it in the study of Systems Chemical Biology, especially for drug discovery. This chapter is organized into two major parts: (1) data representation and integration in the Semantic Web and (2) knowledge discovery in integrated semantic data. Each part begins with a general introduction of techniques, which is followed by a review of several interesting cases, and ends with a discussion of the challenges.

1.2   DATA REPRESENTATION IN THE SEMANTIC WEB

RDF

Resource Description Framework (RDF) is the recommended language for representing data on the Web. It represents data as triples consisting of subjects, predicates, and objects. Every statement can be represented as a set of triples. As Figure 1.2 shows, troglitazone has target PPARG, in which "troglitazone" is a subject, which has "Target" as its predicate, and "PPARG" as its object. Every resource can be identified as a URI (Uniform Resource Identifier). It is recommended that the URI be dereferenceable, meaning we are able to retrieve all the information about the "thing" identified by the URI from the Web: for example, the URI of troglitazone is considered as representing an object, rather than a term, illustrating that the meaning is "traveling" along with the data. If we agree to identify the source as a URI, we are able to link it from one resource to another; and eventually, all the data are linked into a huge graph with semantic meanings. For example, (see Figure 1.2), "troglitazone" is linked to "PPARG" in DrugBank, "PPARG" is linked to "MED1" in HPRD, and "PPARG" is linked to "Diabetes Mellitus" in OMIM. Since the objects of PPARG from HPRD and OMIM share the same identifier, they are thus automatically merged. The names of PPARG that appeared in both databases are represented as the subjects of two triples; given that they state the same "thing," the two triples can be automatically merged.

Sometimes, a triple is not capable of representing a statement with contexts. For instance, paper x reports that "troglitazone" has Target "PPARG," a named graph, usually represented as a URI, is added into the triple to refer to the context (in this case, "paper x").

RDF's data model differs from those two popular data models: the Relational Database and XML. Relational database is a set of tables related by primary keys and foreign keys, XML is a hierarchical tree, and RDF is a graph. In other words, XML and relational database are data models that represent syntactic structures (e.g., a tree structure or a set of tuples) of data; while RDF models data as a set of triples with clearly defined semantics for subjects, predicates, and objects. Therefore, RDF triples form a graph that offers agile management of data. A new triple can be simply added to the existing graph therefore automatically linked to others. As technologies usually generate data with various formats, adopting triples to represent such diverse data would make data management simple.

CONTROLLED VOCABULARY

Triples offer an agile approach to linking data. However, the practice confronts a challenge that originates from the assumption that two objects to be merged should use the same vocabularies. To represent the name, "PPARG," if the predicate in HPRD and in OMIM were using "name" and "proteinName" separately, they would be considered as two different triples, even though they share the same semantic meaning. To avoid this situation, we use common vocabularies to represent the terms. A large number of controlled vocabularies have been developed (see Table 1.1) to normalize the diversity of semantics. If we are willing to develop and adopt more common vocab-

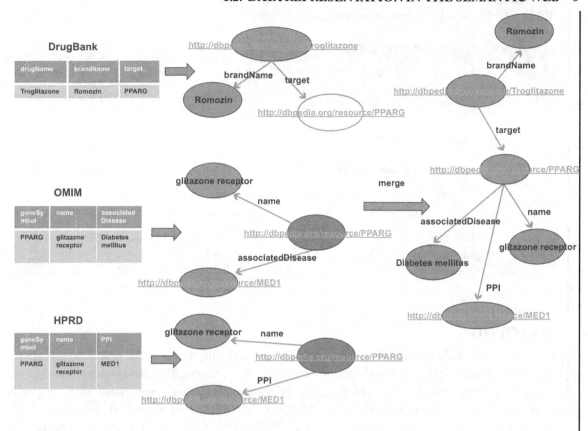

Figure 1.2: RDF demo. Note: Data in relational tables from different sources are converted into RDF triples. As a common vocabulary and URIs are adopted, the triples are linked straightforwardly during the integration.

ularies, the power of linked data will eventually be manifested. For data using different identifiers, an alternative is to use *owl:sameAs* to link them. For example, "troglitazone" is identified as the following URIs in different sources:

- http://dbpedia.org/page/Troglitazone

- http://www4.wiwiss.fu-berlin.de/drugbank/resource/drugs/DB00197

- http://www4.wiwiss.fu-berlin.de/sider/resource/drugs/5591

- http://chem2bio2rdf.org/drugbank/resource/drugbank_drug/DB00197

Table 1.1: Examples of common vocabularies

Name	Namespace	Description
RDFS	http://www.w3.org/2000/01/rdf-schema#	consists of class and property type definitions; Examples: rdfs:label, rdfs:subClassOf
OWL	http://www.w3.org/2002/07/owl#	extends the expressivity of RDFS with additional modeling primitives; Examples: owl: equivalentClass, owl: inverseOf
Dublin Core	http://purl.org/dc/terms/	defines general metadata attributes for articles; Examples: dc:creator, dc:date
Creative Commons	http://creativecommons.org/ns#	defines terms for describing copyright licenses in RDF. Examples: cc:license, cc:permits
foaf	http://xmlns.com/foaf/0.1/	defines terms to link people and information using the Web. Examples: foaf: homepage, foaf:Person
VoID	http://rdfs.org/ns/void#	describes linked datasets. Examples: void: Dataset, void:target

ONTOLOGY

Controlled vocabularies play an important role in normalizing semantics to link data, but are not capable of modeling the relationships beyond *broader/narrower/seeAlso* relationships. An ontology is a formal representation of a set of concepts and their relations (e.g., *is-a* relationship, or specific relations using object properties) so that inference can be enabled. The RDF Schema (RDFS) (http://www.w3.org/TR/rdf-schema) and OWL (Web Ontology Language) provide basic modeling constructs (e.g., *subclass, subproperty, objectProperties*, etc.) to formally describe ontologies. For example, "troglitazone" is a type of "Drug," and "troglitazone" is a *subclass* of "thiazolidinedione." OWL offers a richer vocabulary and extensively expressive power than the RDFS. For example, a "small" molecule is defined as a molecule with a weight of no more than 800 Daltons. Given a new molecule, if its weight is greater than 800 Daltons, it should not be called a "small" molecule. The property "protein-protein interaction" is characterized as symmetric, so the statement, "protein A interacts with protein B" implies that protein B also interacts with protein A. Protégé (http://protege.stanford.edu/) is an open-source tool for developing ontologies that can be represented in RDFS or OWL.

OTHER LANGUAGES

Besides RDFS and OWL, several languages have been developed to assist modeling the complex relations found in life sciences. The Web of data is decentralized. When data are updated, the links that connect them should be renewed as well. For example, if a gene name becomes obsolete or is changed, a file is needed to record the changes. Such issues are referred as *provenance*, which documents basic information about changes, including when, where, what, who, how, and evidence, so as to allow the facts to be reproduced [Zhao et al., 2009].

Biological Expression Language (BEL, http://www.openbel.org/) is a semantic representation language for life sciences that supports both causal and correlative relationships, as well as negative relationships, in the biological system, making it suitable for recording a variety of experimental and clinical findings. BEL is designed to record scientific facts and associated experimental contexts, including information about the biological and experimental system in which the relationships were observed, the cited supporting publications, and the process of curation. For example, abundance of human HIF1A protein hydroxylated at asparagine 803 is represented as: *proteinAbundance (HGNC:HIF1A, proteinModification(H, N, 803))*. It was initially designed and used in 2003 by Selventa (http://selventa.com/) and became open source recently (OpenBel). BEL was not originally represented in RDF, but can be interchangeable with RDF.

Extracting entities and relations from unstructured data is a non-trivial task. An alternative is to encourage publishing results in semantic formats that would allow scientific results to be shared. Nanopublication (http://www.nanopub.org) is an initiative to represent scientific results as a set of publishable and sharable RDF triples, rather than being embedded in a document. One single nanopublication includes an *assertion* and its provenance. An assertion is a minimal unit of thought, expressing a relationship between two concepts. The provenance describes the context of assertion.

1.3   DATA QUERY, MANAGEMENT, AND INTEGRATION

RDFIZATION

Since most data are stored in relational databases, publishing their tables as RDF triples has become a common practice. The RDB2RDF (http://www.w3.org/2001/sw/rdb2rdf/) tools, such as the D2R Server [Bizer and Cyganiak, 2006], use customizable mapping files to convert database contents into RDF triples, and allow the data to be browsed, searched, or dumped in RDF/XML or Turtle formats. The mapping files, usually described as R2RML [Das et al., 2012], are often generated by default, but can be edited, such as, by including an external ontology. Several other wrappers (e.g., Virtuoso http://virtuoso.openlinksw.com/) have been developed to RDFize XML and CSV files. Many public databases have been RDFized, and their source codes are open to the public (https://github.com/bio2rdf).

TRIPLE STORE

A database system specially designed for storing triples is called a *triple store*. Current triple stores include in-memory, native and non-memory, and non-native running on a third-party database. In-memory triple stores (e.g., Jena http://jena.apache.org/) are limited to handling a small set of triples, due to their dependence on memory availability. Most popular triple stores are native and non-memory-based, such as Virtuoso. These triple stores are essentially powered by relational databases. Non-native "virtual" triple stores (e.g., D2R) can manage RDF triples within a relational database. In such a virtual store, there are no real RDF triples, and Sparql queries are transformed into SQL queries, to be executed on the relational database. The triples can be stored in graphDB (e.g., Neo4j, `http://neo4j.org/`) as well. While graphDB does not attempt to support inference, it has a compelling advantage for graph traversal. The benchmark comparison of triple stores for biological data has been discussed and reported that Virtuoso demonstrates balanced performance, compared to several others [Mironov et al., 2012].

SPARQL standards for a SPARQL Protocol And RDF Query Language. This standard has two parts, a query language and a protocol. First, as an RDF query language, like SQL for searching a relational database, SPARQL searches against a graph to retrieve a set of instances that match pre-defined path patterns. The protocol is basically an API that makes it possible to query RDF data by means of an HTTP request, providing a practical way to implement the protocol. RDF vendors often host SPARQL endpoints, where queries are executed. SPARQL can perform queries on a particular service and bind the results with those returned from another service. This functionality enables federated search across multiple SPARQL endpoints [Quilitz and Leser, 2008]. One federated search example of finding the diseases associated with a given drug in two sources is illustrated below:

```
select ?disease where {
SERVICE {http://chem2bio2rdf.org/drugbank/sparql}
{?drug rdf : label "Troglitazone" .
  ?drug drugbank : hasTarget ?target .}
SERVICE {http://chem2bio2rdf.org/omim/sparql}
{ ?target OMIM: associatedDisease ?disease .}
}
```

LINKED OPEN DATA CLOUD

The Semantic Web aims to turn the Web of document into the Web of data, whose entities are machine-readable Web objects. As dereferenceable URIs are used to uniquely identify data, it is possible to link data automatically or semi-automatically. 295 datasets across life sciences and social sciences have been published as linked open data (`http://linkeddata.org/`; see Figure 1.3). It is recommended to follow the "rule of four" in publishing linked data (`http://www.w3.org/DesignIssues/LinkedData.html`).

1. Use URIs as names for things. Use of PURLs (persistent URLs) is suggested.

2. Use HTTP URIs, so that people can look up those names. This means every URI should be dereferenceable. Use either *303 redirect*s or *Hashed URL*s.

3. Provide useful information when someone looks up a URI. Try to use standards (e.g., RDF).

4. Include links to other URIs, so that more things can be discovered. This suggests we should link our data to that of others, and vice versa, so that we can navigate data across different resources.

As a best practice, publishing linked data has requirements of provenance, quality, credit, attribution, and methods, to ensure the reproducibility of results [Bechhofer et al., 2010]. Furthermore, ontology development and mapping are critical in transforming linked data into a knowledge graph. Several guidelines for publishing life sciences data have been proposed [Marshall et al., 2012].

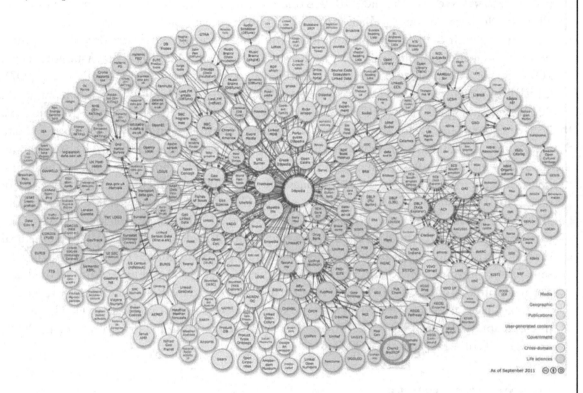

Figure 1.3: Linked open data cloud (`http://linkeddata.org`). Note: The Chem2Bio2RDF bubble was highlighted.

RELATED WORK

A variety of RDF-based Semantic Web resources have been created for biological data and drug data separately. Bio2RDF [Belleau et al., 2008] provides a platform and a strategy for generating and querying of biological RDF data in a distributed framework, with around 4 billion RDF triples across more than 30 biological resources. Linking Open Drug Data (LODD) [Jentzsch et al., 2009], led by the W3C Semantic Web Health Care and Life Sciences Interest Group (HCLS IG), links RDF data from the Linked Clinical Trials dataset derived from ClinicialTrial.gov, DrugBank, and many other sources, with more than 8.4 million RDF triples and 388,000 links to external data sources. Chem2Bio2RDF [Chen et al., 2010b] created an RDF repository for more than 20 public data sources, especially chemogenomics sources, and linked them to each other. OpenPhacts, part of the Innovative Medicines Initiative, created an open pharmacological space using Semantic Web technologies [Williams et al., 2012b]. Other efforts include Linked Life Data (http://linkedlifedata.com), OpenTox (http://www.opentox.org/), and BioRDF (http://www.w3.org/wiki/HCLSIG%20BioRDF%20Subgroup). Thanks to the nature of RDF, all these efforts can be put together, forming a big linked open cloud.

Along with the RDFization, a variety of individual domain ontologies describing entities in Systems Chemical Biology have emerged, for example, Gene Ontology (GO; Ashburner et al., 2000), Sequence Ontology [Eilbeck et al., 2005], PRotein Ontology (PRO; Natale et al., 2007), ChEBI for small molecules [Degtyarenko et al., 2008], Disease Ontology (DO; Schriml et al., 2012), and Biological Pathway Ontology (BioPAX; Demir et al., 2010). Several ontologies have been developed recently to formalize chemical biology experiments and provide guidance for data annotation, including the Minimum Information About a Bioactive Entity (MIABE; Orchard et al., 2011), and the BioAssay Ontology [Visser et al., 2011]. Other domain ontologies, including pharmacogenomics [Dumontier and Villanueva-Rosales, 2009], ligand protein interaction [Choi et al., 2010, Ivchenko et al., 2011], Disease-Drug Correlation Ontology (DDCO; Qu et al., 2009), Translational Medicine Ontology [Luciano et al., 2011], neuromedicine (SWAN; Ciccarese et al., 2008), DDI for drug discovery investigation [Qi et al., 2010], and OBI for biomedical investigations [Brinkman et al., 2010]. These ontologies play an important role in normalizing the semantics of terminologies in life sciences. A number of upper level ontologies, such as the Basic Formal Ontology (BFO, http://www.ifomis.org/bfo/) and the Relation Ontology [Smith et al., 2005], have been proposed, to regulate the generation of domain ontologies to enable broader interoperability. Most of life sciences ontologies are stored in OBO foundry [Smith et al., 2007] or NCBO BioPortal [Noy et al., 2009].

MAJOR CHALLENGES

The greatest and most urgent challenge for data integration is probably data standardization. The techniques of converting data into RDF are not hard, but to build a high quality RDF dataset able to link to others and to be utilized freely without ambiguous semantics is not trivial in practice. Relational databases provide limited data types to restrict data values, but they neither distinguish different entities nor standardize data values using domain terminologies. Therefore, converting relational databases into RDF triples requires the employment of multiple ontologies; we can either create an ontology tailored for the dataset or the application, or utilize other domain ontologies directly. However, choosing an appropriate ontology from the continuously expanding libraries of ontologies is not easy. Although these ontologies are encouraged to be aligned with upper level ontologies, such as the Basic Formal Ontology, these upper level ontologies usually reflect a high level of abstraction which makes their application extremely complicated. For example, the construction of SPARQL is designed to be "friendly," but the diverse ontologies and complicated ontological structures make the formation of queries extremely difficult and counter-intuitive.

The nature of RDF triples likely hampers the modeling of context, which becomes critical in life sciences. Although "named graph" and other approaches have been proposed, none of them have solved the issues. We have to think of a best practice strategy that allows the context to be modeled together with its data, as many current tools and algorithms are designed only for triples and not 4-tuples. For example, one experiment shows drug A interacting with protein B with mutation at position X, so simply using a triple stating that A interacts with protein B is not precise. Should we consider the protein with this mutation as an instance of Class Protein B, or should we consider protein B as an instance of Class Protein and add mutation as a condition of this statement?

Different sets of RDF triples can be merged into a single triple repository, allowing data retrieval, and inference across multiple resources. However, the size of RDF triples can easily reach to billions, and how to index the data appropriately to allow efficient data retrieval remains crucial. An alternative is to use federated search, relying on the HTTP-based SPARQL protocol. However, federated search does not support inference and the integration of the searched results, and efforts are needed to optimize the aggregation of results from different SPARQL endpoints.

1.4   KNOWLEDGE DISCOVERY IN SEMANTICALLY INTEGRATED DATASETS

All data in Systems Chemical Biology can be represented using RDF or OWL to form linked data. In linked data, the same entities can be represented using different URIs, and linked using *owl:sameAs* or *rdfs:seeAlso*. But consolidating these nodes with different URIs representing the same entity to one node requires giving a URI to harmonize the other URIs. This appears trivial, but is hard to realize in a distributed environment. Furthermore, to identify whether two entities are actually the same entity requires in-depth domain knowledge, and remains as an open issue in many disciplines in life sciences.

After merging all nodes representing the same entity into one node and trimming off datatype properties and their values, linked data can be turned into a knowledge graph. A knowledge graph is a heterogeneous network where a node is either a subject or an object of an RDF triple, and an edge is a predicate with rich semantic meaning. After a knowledge graph is formed, graph mining algorithms and inference can be applied.

LOGIC-BASED INFERENCE

Logic-based inference aims to derive new facts from a set of asserted facts, based on a subset of first-order logical propositions created by rule languages. Some basic assertions (e.g., *subclass of*, *subproperty of*, *seeAlso*, etc.) defined in RDF Schema enable simple inference. More expressive languages with restrictions and property descriptions are offered in OWL, where a knowledge base is defined as a set of statements made up of ABOX (i.e., components containing assertions about individuals) and TBOX (i.e., components containing axioms about classes). Reasoning in OWL is widely used in the following situations: (1) consistency check that examines if any contradictory facts exist. For instance, A is a subclass of B and A is a subclass of C; if B and C are stated as disjoint, there will be an inconsistency error; (2) classification that computes the subclass relations between classes to create a class hierarchy; (3) concept satisfiability that checks if it is possible for a class to have any instances; and (4) realization that computes the direct types for individuals. Three profiles of OWL 2 (i.e., OWL 2/EL, OWL 2/QL, and OWL 2/RL; http://www.w3.org/TR/owl2-overview/ have been developed, and each profile is useful in different application scenarios, with respect to reasoning efficiency. For example, OWL 2/EL is useful for an ontology including a large number of properties and classes, while OWL 2/QL favors an ontology involving many instances. In addition to RDF Schema and OWL, rule languages have been developed to facilitate advanced reasoning. The Semantic Web Rule Language (SWRL, http://www.w3.org/Submission/SWRL/) is particularly useful for describing a rule, which includes a Body (antecedent) and a Head (consequent). Whenever the conditions specified in the antecedent hold, the conditions specified in the consequent must hold. Rule Interchange Format (RIF, http://www.w3.org/2005/rules/wiki/RIF_Working_Group) was created for exchanging rules. It categorizes rules into two major sets: (1) logic-based dialects, in which condition and conclusion are monotonic, conclusion only contains logical statements, while change in prediction value is not allowed; and (2) production-rule dialects, which have an *if* part, or condition, and a *then* part, or action. Instead of logical statements in logic-based dialects (http://www.w3.org/TR/2010/REC-rif-prd-20100622/), an action can assert, modify, and retract facts. For example, a rule to infer compound and disease relation can be written as [Zhu et al., 2011]:

```
[(?CompoundID WO:isActiveIn ?Bioassay) ,
(?Bioassay WO:hasGene ?Gene),
(?Gene WO:isAssociatedWith ?Disease)
-> (?CompoundID WO:mightHasDisease ?Disease)]
```

It is interpreted as: if a compound is active in a bioassay that links to a target, and this target is associated with a disease, then the compound might be associated with the disease. The rules can be executed by several reasoners, for example, Pellet [Sirin et al., 2007], FaCT++ [Tsarkov and Horrocks, 2006], RACER [Haarslev and Möller, 2001]), and HermiT [Shearer et al., 2008].

Reasoning can also be executed in ontology editors such as Protégé or via APIs (e.g., OWLAPI, Horridge and Bechhofer, 2008). Dentler et al. [2011] provide a systematic comparison of different reasoners.

SEMANTIC GRAPH MINING

Semantic graph mining is specified to the integrative modeling of the network by analyzing the topology and semantics of its components. Analysis of the topology of the network has been widely applied, especially when the large scale-free nature of many real networks has been observed. For example, analysis of protein-protein interaction networks helped identify lethal genes of biological systems [Jeong et al., 2001]; analysis of drug target networks led to the conclusion that a drug tends to interact with multiple targets, rather than just one [Yıldırım et al., 2007]; and bipartite drug target networks can infer unknown drug target pairs [Bleakley and Yamanishi, 2009]. Graph mining algorithms have been developed to analyze social behaviors, where nodes usually are friends, authors, and conferences [Han et al., 2012]. Semantic graph mining can be used to conduct similarity and association assessment, ranking, clustering, and community detection [Chen et al., 2009b]. The key is to assess the relation between two homogeneous nodes (referred to as "similarity"), or two heterogeneous nodes (referred to as "association"), which can be further extended to perform ranking, clustering, and other advanced tasks.

(1) Similarity: In a homogeneous network, similar nodes are clustered solely by their connectivity and the weight of their edges. However, as the nodes and edges have semantic meanings in heterogeneous graphs, similarity can be assessed in a context. For example, drug similarity could be assessed either by the properties of the nodes related to structural properties (e.g., substructures) if structure similarity is desired, or by the properties of the nodes related to its targets and their therapeutic indications if functional similarity is desired. Several similarity measures [Liben-Nowell and Kleinberg, 2007] have been developed based on the properties of neighborhood (e.g., the number of common neighbors), the ensemble of paths (e.g., the length of shortest paths, PageRank [Page et al., 1999], SimRank [Jeh and Widom, 2002]), and other approaches using semantics (e.g., SemRank [Anyanwu et al., 2005], Path Sim [Sun et al., 2011]). Some algorithms are actually equivalent to other classic algorithms, for instance, Jaccard index derived from the number of common neighbors is essentially the same as Tanimoto coefficient.

(2) Association: Association measures the relation between any two entities; sometimes it is called link prediction, predicting whether a link between two nodes is likely to be established. Some of the relations (e.g., drug target interactions) have not been tested experimentally, but they can be fished out by the analysis of network-based factors in existing data [Chen et al., 2012b]. The idea assumes that two nodes are associated only if they are linked directly or indirectly, and the association can be assessed by the topology and semantics of the subgraph between two objects. As shown in Figure 1.4(a), the association in the first path is weaker than in the second one, as the two compounds in the second one share a more distinctive substructure than in the first. The

distinctiveness is determined by the degree of the node. Meanwhile, the association of the two paths in Figure 1.4(b) varies due to different semantic meanings.

Aleman-Meza et al. 2005 proposed a framework to rank relations by combining semantic and statistical metrics. A number of algorithms [Chen et al., 2012b, Eronen and Toivonen, 2012, Liekens et al., 2011] transform association to probabilities that traverse from one node to another via paths. The paths are considered as features, and are trained using conventional machine learning algorithms. Some of the path similarity measurements in heterogeneous networks can be transformed to assess association as well [Sun et al., 2011].

Compared with many machine learning algorithms, semantic graph mining possesses at least two compelling advantages. Many supervised algorithms only render a probability, and treat the whole learning (or reasoning) process as a "black box" that is hard to conceive. But due to the nature of graphs, semantic graph mining is capable of visualizing relations and illustrating them explicitly. In addition, semantic graph mining that considers both the topology and the semantics of the graph makes it possible to leverage multi-dimensional factors related to the relations much easier than other methods designed to capture only one type of information. For example, several works on drug target prediction only focus on an individual piece of information, for example, substructure, side effect, gene expression, or chemical ontology [Chen et al., 2012a].

RELATED WORK

Many ontologies have been used in biological analysis, for instance, comparison of gene expression profiles based on annotated gene ontology (GO) terms [Khatri et al., 2012]. Reasoning on ontology is often used during the cycle of ontology development, for example, to validate a new term assigned to the ontology. Reasoning across different domain ontologies makes it possible to explore new relations. Hoehndorf et al. [2012] identified disease pathways using multi-ontology enrichment analysis by integration of a chemical ontology, a disease ontology, and pathway data. Reasoning upon GO, the Human Disease Ontology, and the Mammalian Phenotype Ontology provided an effective way to answer biologically rich questions [Jupp et al., 2012]. Ontologies also have been used for small molecule classification and lipid annotation [Chepelev et al., 2011], and for protein classification [Wolstencroft et al., 2006]. However, automated reasoning upon a large-scale dataset is limited, as the semantics of relations and entities in many biomedical ontologies have not been formally established; aligning with upper level ontologies (e.g., BFO) would help manage contradictory class definitions and make ontologies more interoperable [Chepelev et al., 2011].

Semantic graph mining based on heterogeneous networks has been widely applied. Bio-Graph [Liekens et al., 2011] integrated 21 databases pertaining to genes, diseases, compounds, pathways, ontology terms, protein domains, diseases, gene families, and microRNAs, and developed a probabilistic model to prioritize disease genes. It computed a global priori importance of each concept using *infinite random walk*, and then computed the a posteriori probability of visiting each target concept (e.g., a disease) from the source target (e.g., a gene) by *random walk*

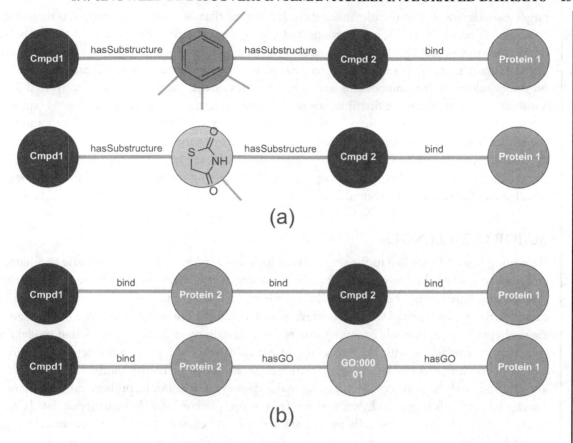

Figure 1.4: Topology and semantics are important in association. Both (a) topology and (b) semantics contribute to association.

with restarts. Each concept (e.g., a gene) was scored with respect to the target concept (i.e., disease) based on the combination of the two types of probabilities, and their probable paths were identified by backtracking heuristics. For example, the PRL gene is found to be associated with schizophrenia, supported by the evidence which schizophrenia drugs (i.e., aripiprazole and risperidone) are able to change the expression of PRL. BioMine [Eronen and Toivonen, 2012] used a large-scale heterogeneous network to rank disease genes in a different way. It was assumed that if a candidate gene is proximal to the disease genes in the network, it might be related to the disease. To measure proximity, it explored various methods, including the probability of best path, *random walk*, and several others. It weighed the edge by the combination of relevance, informativeness, and reliability. The performance is highly dependent on the weighting schema, which requires manual adjustments. SLAP [Chen et al., 2012b] was developed to assess drug target association using semantic linked data. It assumed that two objects are related if they share common objects or their indirectly connected objects are related. It first retrieved all the paths between source object (e.g., drug) and target object (e.g., target) and categorized the paths into different path patterns based on path semantics. It computed a raw score for each path based on the topology of the components and the semantics of the connections, and converted it to a z score based on the score distributions of the path patterns. The association score is equal to the sum of the z scores that are greater than zero, and its significance is assessed by a random model, indicating the probability of observing the score by chance. Figure 1.5 shows an example of linking troglitazone to PPARG. Given that some links are spurious (e.g., false positives) in the graph, they may be identified retrospectively by predicting the probability of the links being established [Eronen and Toivonen, 2012].

MAJOR CHALLENGES

Reasoning is widely applied in the stage of ontology development to assist automatic annotation and quality control. However, cross domain reasoning is still very limited. Besides the fact that domain ontology has not been made fully interoperable, it is hard to precisely model the complicated relations that are valid within a context or with certainty. For instance, the drug troglitazone causes hypertension, but only in some patients under certain conditions. The rules that model the drug and its association with the disease might not be capable of capturing the actual scenario.

Incomplete and low quality data restrict the accuracy of predictive models. Spurious data and contradictory facts are common among major datasets. For example, protein-protein interaction based methods (e.g., two hybrids) suffer from incompleteness and low quality of data [Chen et al., 2009c]; bioassays, especially in primary screening [Schierz, 2009], often generate a large number of false positives; structures in many chemistry databases are assigned incorrectly, mainly due to the lack of "gold" standards [Williams et al., 2012a]; and a large number of GO annotations are predicted without manual inspection [Khatri et al., 2012].

Edges should be weighted appropriately, based on their reliability and importance to the association and its likelihood. But many current algorithms rely on empirical adjustment or prior domain knowledge, limiting their generalization. In addition, networks in life sciences, in contrast to those in social networks, include a large number of negative links (e.g., a compound is inactive against a protein) that should be adjusted in the model. Algorithms like *random walk* need to be modified accordingly.

1.5   Chem2Bio2RDF

Chem2Bio2RDF is a project to explore the major Semantic Web techniques and apply them to the area of Systems Chemical Biology.

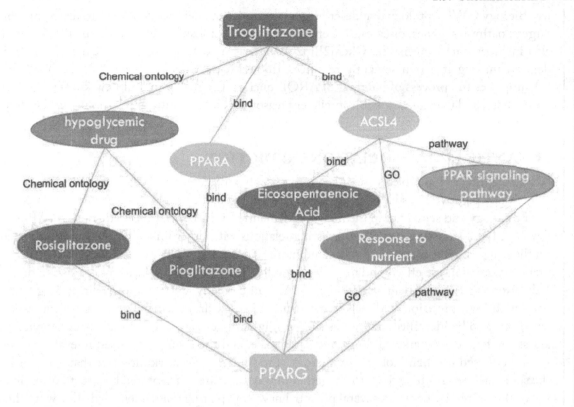

Figure 1.5: Paths between troglitazone and PPARG.

DATA REPRESENTATION AND INTEGRATION USING RDF

Link Open Data Initiatives, such as Bio2RDF and LODD, aim to link biological data and drug data, respectively, using RDF. But the effort of linking data in Systems Chemical Biology that crosses the domains of chemistry and biology is very limited. In this project, a single RDF repository was created by aggregating data from multiple public data sources that are cross-linked to Bio2RDF and LODD, which allows performing cross-data source query using SPARQL. The utility of Chem2Bio2RDF was demonstrated in investigating polypharmacology, and identifying potential multiple pathway inhibitors.

DATA REPRESENTATION AND INTEGRATION USING OWL

Since current RDFization tends not to employ ontology, its utility could be greatly improved if an ontology is applied for annotating and consolidating data which enables a richer range of semantic queries. Chem2Bio2OWL was developed as a generalized chemogenomics and Systems Chemical Biology OWL ontology that describes the semantics of chemical compounds, drugs, protein targets, pathways, genes, diseases, side effects, and the relationships between them. The ontology also includes data provenance. Chem2Bio2OWL was used to annotate the Chem2Bio2RDF dataset, making it a rich semantic resource. Through case studies, we demonstrate how this (1) simplifies the process of building SPARQL queries, (2) enables useful new kinds of queries on the data, and (3) makes possible intelligent reasoning and semantic graph mining in the study of chemogenomics and Systems Chemical Biology.

SEMANTIC LINK ASSOCIATION PREDICTION

After data integration, objects are linked into a graph, with nodes and edges annotated semantically. Many putative links have not been established, but some of them can be inferred based on their topology and semantics in the network. Semantic Link Association Prediction (SLAP) was developed as a statistical model for assessing relations with other linked objects, and was tested on the drug-target association. Understanding the interaction of drugs with multiple targets can identify potential side effects and toxicities, as well as possible new applications of existing drugs. Validation experiments demonstrate that SLAP can correctly identify known direct drug target pairs with high precision. Indirect drug target pairs (e.g., drugs which change gene expression level) can also be identified, but not as successfully as direct pairs. SLAP was further applied to assess the targets of marketed drugs on a large scale, and the results were used to evaluate drug similarity based on their biological functions. The similarity network indicates that drugs from the same disease area tend to cluster together in ways that are not captured by structural similarity. In this similarity network, several potential new drug pairings were identified. This work thus provides a novel, validated alternative to existing drug target prediction algorithms.

CHAPTER 2

Data Representation and Integration Using RDF

2.1 BACKGROUND

Recent advances in chemical and biological sciences have led to an explosion of new data sources about genes, proteins, genetic variations, chemical compounds, diseases and drugs. Through integrated and intelligent data mining, this information could provide important insights into the complex functions of biological systems and the actions of chemical compounds or drugs on these biological systems. However, this can only be achieved when data is semantically integrated (i.e., using multiple data sources that are connected in meaningful ways), and in particular when chemical and biological resources are integrated [Slater et al., 2008, Wild, 2009].

There are critical problems in biology that can only be answered through computational analysis of this kind of integrated chemical and biological information. For example, it is considered increasingly important to profile existing and potential new drugs for their effects across many protein targets, not just a single target of interest, this kind of profiling is known as *polypharmacology* [Hopkins, 2008]. Only by exploring the relationships of the drugs to all the target information that is available can we determine this profile. Further, the polypharmacologic actions of drugs on targets that fall within the same pathway can determine the drug's ability to interrupt pathways at multiple points, and thus provide greater efficacy. Relationships between these pathways and potential side effects of drugs, or chemicals that are being considered as drugs (such as undesirable inhibition of a pathway), can only be determined by large-scale analysis of the impact of the chemicals on known pathway systems [Scheiber et al., 2009]. The need to address such problems has led to the emergence of the field of Systems Chemical Biology [Oprea et al., 2007], a field that covers the computational analysis of integrated chemical and biological information for the enhancement of biological understanding.

Implementing such an integrated system involves the creation of large networks of linked compounds, protein targets, genes, pathways, drugs, diseases, and side effects from multiple heterogeneous sources. It must be possible to query these data in ways that go beyond querying of a single source, and allow inferences on crossed domains. For example, a positive experimental test of a chemical compound in a biological enzymatic assay where the enzyme is associated with a particular metabolic pathway constitutes a probable action of that compound on the pathway. Currently, there are significant barriers to carrying out this kind of analysis. Many of the needed data sources overlap and cover similar data (i.e., we refer to them as homogeneous or semi-homogeneous data sources), but with slightly different focuses. All data sources tend to be published in diverse formats (e.g., text files, CSV, PDF, XML, relational databases, and so on) which are structured or unstructured. The semantic relationship of these datasets to each other is often unclear.

Recent Semantic Web technologies provide a way to integrate heterogeneous data. The Semantic Web [Berners-Lee et al., 2001] has demonstrated its utility in life sciences, healthcare, and drug discovery [Chen et al., 2009a, Neumann, 2005]. Various semantic languages have been established to represent and query the semantic meaning of data and relationships. Here, RDF was used with a unified data model to model chemogenomics and Systems Chemical Biology data, and SPARQL for querying these data.

A variety of RDF-based Semantic Web resources have already been created for biological data and drug data separately, for instance, Bio2RDF [Belleau et al., 2008], Linking Open Drug Data (LODD, Jentzsch et al., 2009), YeastHub [Cheung et al., 2005], BioDash [Neumann and Quan, 2006], and BioGateway [Antezana et al., 2009].

2.2 METHODS

DATASETS

The datasets can be organized into six categories, based on the kinds of biological and chemical concepts they contain. These categories are: chemical and drug, protein and gene, chemogenomics, systems (i.e., PPI and pathway), phenotype (i.e., disease and side effects), and literature. Some data sources are listed in multiple categories. As the RDF representations of the dataset in other domains have been reported (i.e., LODD, Bio2RDF), our focus is on the chemogenomics data. Some major data sources are provided below.

1. PubChem BioAssay [Wang et al., 2009]: PubChem is a public repository of chemical information including structures of small molecules and various molecular properties. It has three databases namely Compound, Substance, and BioAssay. The BioAssay database contains experimental results of the compounds in PubChem that have been tested in MLI screening centers or elsewhere against particular biological targets. We selected only enzymatic assays that study a single target.

2. KEGG [Kanehisa et al., 2006]: KEGG is a collection of online databases that curate chemical, genome, and pathway information. One of its databases, KEGG Ligand, has biological molecules associated with their enzymes.

3. CTD [Davis et al., 2011]: Comparative Toxicogenomics Database (CTD) focuses on the effects of environmental chemicals on human diseases.

4. BindingDB [Liu et al., 2007]: BindingDB provides binding affinities on the interactions of proteins with small, drug-like molecules. The data are extracted from scientific literature, focusing on the proteins that are drug targets or candidate drug targets.

5. MATADOR [Günther et al., 2008]: Unlike DrugBank, Manually Annotated Targets and Drugs Online Resource (MATADOR) provides indirect interactions between chemicals and targets that are collected by automated text mining approaches, followed by manual curation.

6. QSAR: We manually collected a set of small QSAR datasets from two websites `http://www.cheminformatics.org/`, and `http://www.qsarworld.com/qsar-datasets.php`, which aggregate all the QSAR datasets published in the literature. Only datasets with explicit targets were selected.

7. TTD [Zhu et al., 2010]: Therapeutic Target Database (TTD) provides information about known therapeutic proteins and nucleic acids described in the literature, and their corresponding drugs/ligands.

8. DrugBank [Wishart et al., 2006]: Drugbank combines detailed drug (i.e., chemical, pharmacological, and pharmaceutical) data with comprehensive drug target (i.e., sequence, structure, and pathway) information. The database contains 6,711 drug entries and 4,227 non-redundant proteins.

9. ChEMBL [Gaulton et al., 2012]: ChEMBL focuses on interactions and functional effects of small molecules binding to their macromolecular targets. It provides 500,000 bioactive compounds, their quantitative properties, and bioactivities (e.g., binding constants, pharmacology, ADMET, etc.).

10. PDSP (`http://pdsp.med.unc.edu/`): The PDSP Ki database provides information on abilities of drugs to interact with an expanding number of molecular targets. It provides a Ki value (i.e., one measure of binding affinity) for each of its 766,000 interactions.

11. PharmGKB [Klein et al., 2001]: PharmGKB curates primary genotype and phenotype data, annotating gene variants and gene-drug-disease relationships published in the literature.

12. Binding MOAD [Benson et al., 2008]: The "Mother Of All Databases" (MOAD) collects all well-resolved protein crystal structures with clearly identified biologically relevant ligands annotated with experimentally determined binding data extracted from the literature. All structures are extracted from the Protein Data Bank (PDB), and contain high-quality ligand-protein binding data.

SYSTEMS

Conversion to RDF is shown in Figure 2.1. Some of the datasets were already stored in relational databases in prior works [Chen and Wild, 2010, Chen et al., 2009a], and they were simply converted into RDF using the D2R server [Bizer and Cyganiak, 2006]. Otherwise, we acquired the raw datasets (i.e., by downloading from websites), stored them into our relational database using

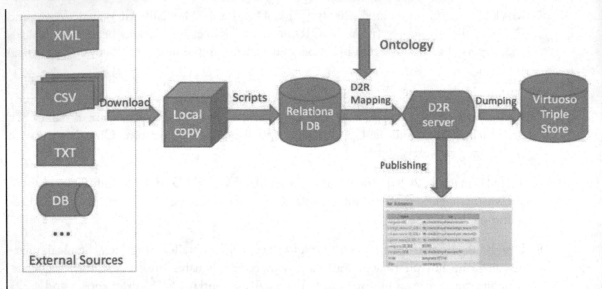

Figure 2.1: RDFization workflow.

customized scripts, and published them as RDF triples using the D2R server (see Figure 2.2). The core part is the mapping file in which tables, columns, and values are mapped to classes, predicates, and objects, respectively. Each entry in a table is transformed into a RDF statement, and its primary key is often used to compose the URI of the subject. D2R could generate a default mapping file, but since the database schema is different from the RDF model, we manually configured the mapping file to make sure that every element is semantically well-annotated and linked to other sources if possible. For instance, join tables are meaningless; some column names can be replaced by terms from common vocabularies; and compounds in DrugBank should be linked to those in PubChem. In such a way, we converted each database into RDF triples and uploaded them into a Virtuoso triple store. In the triple store, a named graph is used to label each database. In order to enable search efficiency, a query should specify a named graph to prevent it from searching against the entire triple store.

Part of the raw data was processed following the "best-practice" guidelines. For example, BindingDB uses a string (i.e., "> 0.5") instead of a number to present its binding affinity. If the user wants to select data greater than 0.4, the text does not support this kind of search, so we manually separated the text into two parts, namely "operator" and "number." Many chemical identifiers (e.g., SMILES, SDF, InChi, and CID number) are available to represent the same chemical, but the machine does not know they actually refer to the same subject. Thus, we assigned each chemical compound a CID number (e.g., PubChem Compound ID) if available, as PubChem is a hub of public compounds. Meanwhile, since proteins can be represented by their GI numbers, UniProt IDs, EC numbers, PDB IDs, and gene symbols, we kept their original

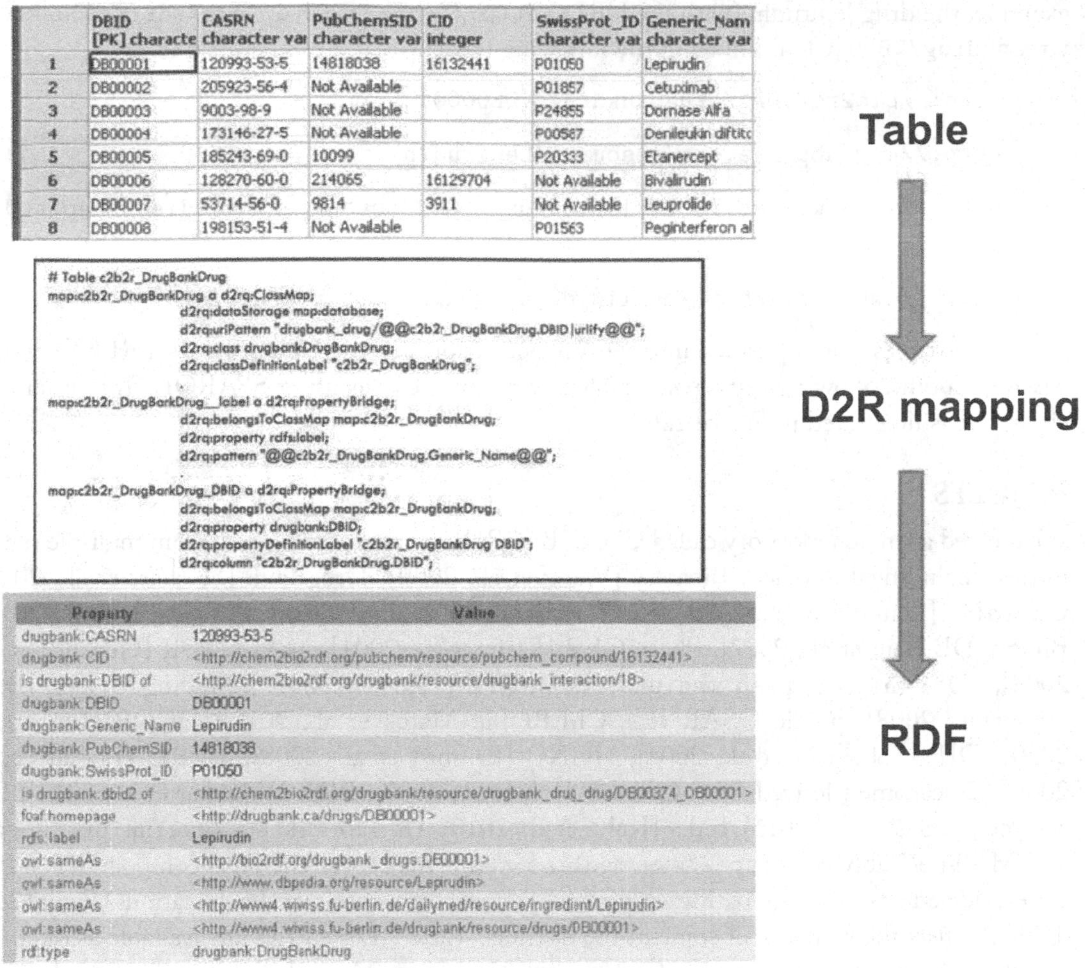

	DBID [PK] characte	CASRN character var	PubChemSID character var	CID integer	SwissProt_ID character var	Generic_Nam character var
1	DB00001	120993-53-5	14818038	16132441	P01050	Lepirudin
2	DB00002	205923-56-4	Not Available		P01857	Cetuximab
3	DB00003	9003-98-9	Not Available		P24855	Dornase Alfa
4	DB00004	173146-27-5	Not Available		P00587	Denileukin diftito
5	DB00005	185243-69-0	10099		P20333	Etanercept
6	DB00006	128270-60-0	214065	16129704	Not Available	Bivalirudin
7	DB00007	53714-56-0	9814	3911	Not Available	Leuprolide
8	DB00008	198153-51-4	Not Available		P01563	Peginterferon al

Table

```
# Table c2b2r_DrugBankDrug
map:c2b2r_DrugBankDrug a d2rq:ClassMap;
        d2rq:dataStorage map:database;
        d2rq:uriPattern "drugbank_drug/@@c2b2r_DrugBankDrug.DBID|urlify@@";
        d2rq:class drugbank:DrugBankDrug;
        d2rq:classDefinitionLabel "c2b2r_DrugBankDrug";

map:c2b2r_DrugBankDrug__label a d2rq:PropertyBridge;
        d2rq:belongsToClassMap map:c2b2r_DrugBankDrug;
        d2rq:property rdfs:label;
        d2rq:pattern "@@c2b2r_DrugBankDrug.Generic_Name@@";

map:c2b2r_DrugBankDrug_DBID a d2rq:PropertyBridge;
        d2rq:belongsToClassMap map:c2b2r_DrugBankDrug;
        d2rq:property drugbank:DBID;
        d2rq:propertyDefinitionLabel "c2b2r_DrugBankDrug DBID";
        d2rq:column "c2b2r_DrugBankDrug.DBID";
```

D2R mapping

Property	Value
drugbank:CASRN	120993-53-5
drugbank:CID	<http://chem2bio2rdf.org/pubchem/resource/pubchem_compound/16132441>
is drugbank:DBID of	<http://chem2bio2rdf.org/drugbank/resource/drugbank_interaction/18>
drugbank:DBID	DB00001
drugbank:Generic_Name	Lepirudin
drugbank:PubChemSID	14818038
drugbank:SwissProt_ID	P01050
is drugbank:dbid2 of	<http://chem2bio2rdf.org/drugbank/resource/drugbank_drug_drug/DB00374_DB00001>
foaf:homepage	<http://drugbank.ca/drugs/DB00001>
rdfs:label	Lepirudin
owl:sameAs	<http://bio2rdf.org/drugbank_drugs:DB00001>
owl:sameAs	<http://www.dbpedia.org/resource/Lepirudin>
owl:sameAs	<http://www4.wiwiss.fu-berlin.de/dailymed/resource/ingredient/Lepirudin>
owl:sameAs	<http://www4.wiwiss.fu-berlin.de/drugbank/resource/drugs/DB00001>
rdf:type	drugbank:DrugBankDrug

RDF

Figure 2.2: Demo of D2R for converting tables into RDF triples.

protein identifiers in the datasets and linked them to UniProt IDs, which were further linked to pathways, protein-protein interactions, and diseases.

We adopted PubChem Compound ID (CID) and UniProt ID as the identifier for compounds and protein targets separately. The compounds represented by other data formats (e.g., SMILES, InChi, and SDF) were mapped to CID via InChi keys. We linked our data to LODD and Bio2RDF using *owl:sameAs*. Since LODD and BioRDF have strict namespace definition and dereferenceable URIs, it is straightforward to link them through a D2R mapping file. For example, the drug lepirudin (with the URI as `http://chem2bio2rdf.org/rdf/resource/dr ugbankdrug/DB00001`) is linked to the following URIs:

`http://bio2rdf.org/drugbankdrugs:DB00001,`

`http://www.dbpedia.org/resource/Lepirudin,`

`http://www4.wiwiss.fu-berlin.de/dailymed/resource/ingredient/Lepirud in,` and

`http://www4.wiwiss.fu-berlin.de/drugbank/resource/drugs/DB00001`

The triples were uploaded into the Virtuoso triple store, which provides a REST service-based endpoint, allowing third-party tools to query the triples using SPARQL. The system architecture is illustrated in Figure 2.2.

RESULTS

We created a single repository called Chem2Bio2RDF by aggregating data from multiple repositories, including PubChem Bioassay [Wang et al., 2009], DrugBank [Wishart et al., 2006], ChEMBL [Gaulton et al., 2012], KEGG [Kanehisa et al., 2006], CTD [Davis et al., 2011], BindingDB [Liu et al., 2007], PharmGKB [Klein et al., 2001], MATADOR [Günther et al., 2008], PDSP (`http://pdsp.med.unc.edu/`), TTD [Zhu et al., 2010], Binding MOAD [Benson et al., 2008], public QSAR sets, ChEBI [de Matos et al., 2010], DCDB [Liu et al., 2010], DIP [Salwinski et al., 2004], HGNC [Bruford et al., 2008], HPRD [Prasad et al., 2009], Diseasome (derived from OMIM, Goh et al., 2007), PDB [Velankar et al., 2012], Reactome [D'Eustachio, 2011], Sider [Kuhn et al., 2010], UniProt (`http://www.uniprot.org/`) and Medline/PubMed (`http://www.medline.com/`) that cover chemical compounds, targets, genes, side effects, diseases, pathways, protein-protein interactions, and publications (Figure 2.3). Table 2.1 lists the statistics of some of the datasets. Their RDF entries are available at `http:// chem2bio2rdf.wikispaces.com/individual+sources`. More than 100 million triples were generated and uploaded into the Virtuoso triple store (`http://cheminfov.informatics.indi ana.edu:8890/sparql`), where queries are executed (see Figure 2.4).

Note: Each node presents one data source, and two data sources are linked if they share common information. Node size and edge width are determined by the number of triples and the number of common data points respectively. Nodes are colored by their RDF vendors (e.g., Bio2RDF, LODD) and edges are colored by their linkage type. For instance, PubChem is linked to ChEMBL via PubChem CID, and ChEMBL is linked to ChEBI via ChEBI ID.

CASE STUDY 1: LINKING DRUGBANK AND PUBCHEM TO INVESTIGATE DEXAMETHASONE POLYPHARMACOLOGY

Since approximately 35% of known drugs have more than one target, the efficacy of many drugs is increasingly thought to come from their effect on multiple targets. The study of drugs with

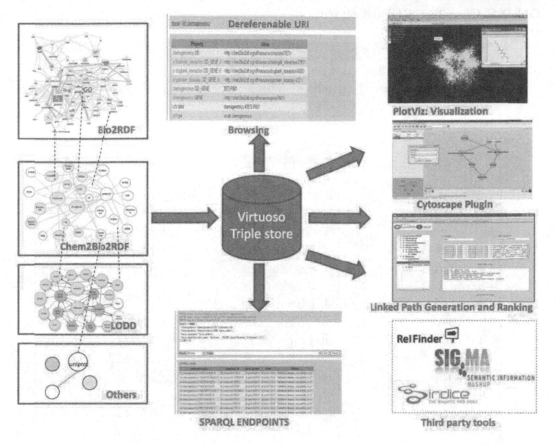

Figure 2.3: Chem2Bio2RDF architecture.

multiple targets is known as polypharmacology. We recently studied the utility of data in Pub-Chem for identifying cases of polypharmacology [Chen et al., 2009a], as well as how chemical and biological data can be mined on a large scale [Wang et al., 2007a]. We can now extend this research, using Chem2Bio2RDF, to incorporate data from DrugBank, as well as PubChem. In particular, if a compound has the same multiple targets as a marketed drug, but has a different chemical structure, it could be a candidate for a novel new therapy. Conversely, if we have already established polypharmacologic action of particular known drugs, can we find other interesting drug-like compounds that show similar polypharmacology? These questions can be formulated as a query: "Find all the drug-like compounds in PubChem BioAssay that share at least two targets with a drug in DrugBank." We can now translate this into a SPARQL query of Chem2Bio2RDF (in this example, using dexamethasone, an anti-inflammatory 9-fluoro-glucocorticoid that interacts with six proteins, as the drug of interest).

Table 2.1: Chem2Bio2RDF dataset statistics

Data Source	RDF Resource Name	# of RDF Triples
PubChem Compound	compound	233,852
PubChem BioAssay	pubchem bioassay	1,715,247
ChEBI	chebi	2,237,330
KEGG	kegg_ligand	96,000
KEGG	kegg_interaction	70,029
KEGG	kegg_pathway_protein	84,760
CTD	ctd_interaction	2,443,826
CTD	ctd_chem_disease	2,025,513
BindingDB	bindingdb_ligand	223,818
BindingDB	bindingdb_interaction	800,016
PharmGKB	pharmgkb_drugs	14,760
PharmGKB	pharmgkb_genes	340,808
PharmGKB	pharmgkb_rela ions	73,276
DrugBank	drugbank_drug	47,640
DrugBank	drugbank _interaction	111,001
UniProt	uniprot	34,951
HPRD	hprd	408,177
Reactomc	reactome	21,985
DIP	dip	1,113,840
OMIM	omim	23,432
SIDER	sider	305,510
PubMed	pubmed2compound	269,178

The query starts with retrieving the active compounds, followed by the identification of targets, which are then linked to drugs in DrugBank. In PubChem BioAssay, an outcome represents a binary result (e.g., "1" for "inactive," "2" for "active"), and the normalized score measures the activity. We select the compounds with an activity score greater than 50. In addition, it is expected that retrieved compounds are drug-like; the function *ruleofFive*—derived from *Lipinski's Rule of Five*, a rule of thumb to evaluate druglikeness—enables us to filter out the compounds that do not pass the five rules. One path is then created if the compound is able to link to the input drug (i.e., dexamethasone) by sharing one common target. However, only a compound that has at least two paths reaching the input drug shows polypharmacology; thus, we group the paths, and select the compounds with more than two link paths as the output. This query process is illustrated in Figure 2.5.

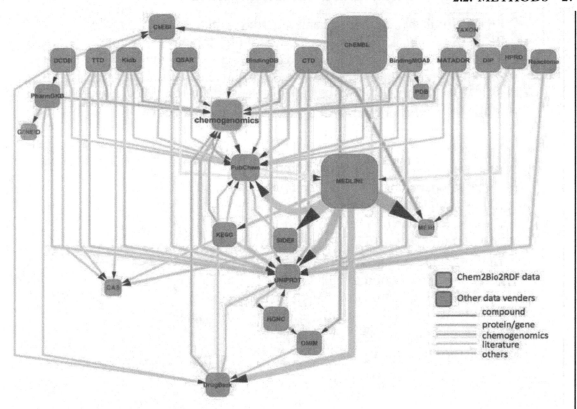

Figure 2.4: Chem2Bio2RDF dataset.

Nine of the retrieved active compounds are active against at least two of the same protein targets, all of which are drug-like (i.e., in terms of *Lipinski's Rule of Five*). These compounds make sense from a medicinal chemistry perspective. For example, dexamethasone and one result, tocris-1126 (CID: 6603742), have similar activity in NFKB1 and NR3C1, because they only have a slight difference in their stereochemistry. The activity of dexamethasone is also similar to that of another search hit, hydrocortisone (CID: 5754), as the addition of the methyl and fluorine to hydrocortisone has no effect on the activity, but improves the compound's druglikeness. In our generalized mapping process, we found 55 significant proteins in DrugBank that are studied in PubChem BioAssay. 27 drugs have corresponding active compounds showing polypharmacology. The SPARQL query is shown as follows:

```
PREFIX pubchem: <http://chem2bio2rdf.org/pubchem/resource/>
PREFIX drugbank: <http://chem2bio2rdf.org/drugbank/resource/>
PREFIX uniprot: <http://chem2bio2rdf.org/uniprot/resource/>
SELECT ?compound_cid (count(?compound_cid) as ?active_assays)
FROM <http://chem2bio2rdf .org/pubchem>
FROM <http://chem2bio2rdf .org/drugbank>
FROM <http://chem2bio2rdf .org/uniprot>
WHERE {
?bioassay pubchem:CID ?compound_cid .
?bioassay pubchem:outcome ?activity . FILTER (?activity=2) .
?bioassay pubchem:Score ?score . FILTER (?score>50) .
?bioassay pubchem:gi ?gi .
?uniprot uniprot:gi ?gi .
?uniprot uniprot:geneSymbol ?gene .
?drugbank_interaction drugbank:gene ?gene .
?drugbank_interaction drugbank:DBID ?drugbank_drug .
?drugbank_drug drugbank:GenericName ?drug_name .
FILTER (?drug_name=''Dexamethasone") .
   } GROUP BY ?compound_cid HAVING (count(*)>1)
```

CASE STUDY 2: LINKING KEGG/REACTOME PATHWAYS AND PUBCHEM TO IDENTIFY POTENTIAL MULTIPLE-PATHWAY INHIBITORS FOR MAPK

Traditional drug discovery approaches focus on identifying a potential target in a disease-related biological pathway, and then finding a drug molecule to interact with this target. However, divergent and redundant pathways in humans often enable a system to keep functioning if one pathway is blocked, as there is an alternative pathway to compensate [Keith et al., 2005]. This can get quite complex, as illustrated in Figure 2.6, where it is inappropriate to inhibit the upstream node A, which has a downstream node B that performs other biological functions. Therefore, in order to block the whole pathway, the drug has to inhibit targets C and D, which are located in separate branches. If the compound in PubChem is actively against C and D, it might be of interest for further investigation if it has efficacy toward the disease caused by the dysfunction of the pathway. We can therefore begin to identify such compounds with the query: "Find all the compounds in PubChem that are active towards at least two targets that are in a given pathway." We can formalize this into a Chem2Bio2RDF query; first, we generate a rule linking compounds with pathways via UniProt. This rule can be illustrated as follows.

The rule generates triples consisting of compound and pathway, which are further refined by its activity (outcome is 2) and pathway name (including the MAPK signaling pathway). Finally,

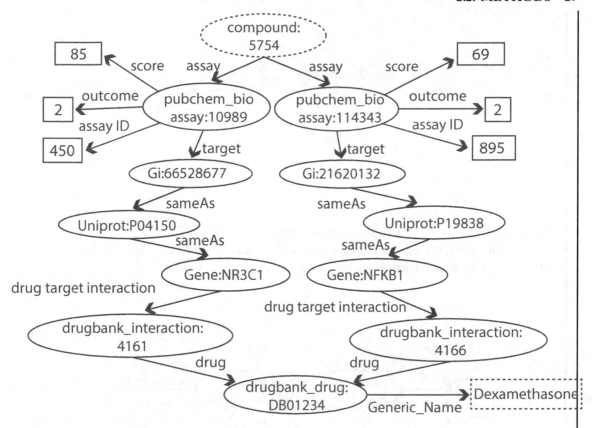

Figure 2.5: Graphical representation of the SPARQL query in the polypharmacology study. Note: PubChem compounds (e.g., CID 5754) are identified that are active in bioassays and are associated with protein targets, genes (via UniProt), and drugs (via DrugBank). The resultant compounds therefore have a similar activity profile to dexamethasone.

like the linking in case study 1, the results are grouped, and only compounds that are multiple pathway inhibitors are selected (see Figure 2.7).

The MAPK signaling pathway plays an important role in coordinating cell proliferation, differentiation, and death. The inhibitors of proteins involved in the pathway are widely studied, but the robustness of this pathway leads to drug resistance. Cisplatin, for example, is used to treat ovarian cancer but the development of resistant cell populations limits its efficacy in long-term trials. It has been suggested that targeting ERK-MKP-1 could destroy this pathway and further overcome cisplatin resistance in human ovarian cancer treatment [Wang et al., 2007b]. One compound (CID: 573747) was found in the retrieved results that has never been reported in the literature, but which can apparently inhibit both ERK2 and MKP-1. We might consider this

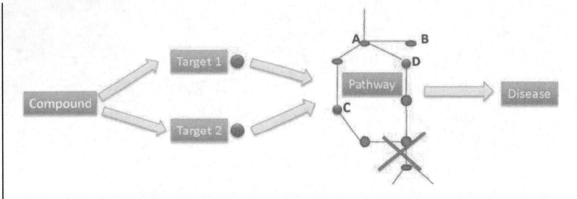

Figure 2.6: Illustration of polypharmacology in pathways. Note: The compound is actively against two proteins which are located in two branches of a pathway associated with one disease. Targeting either node C or node D does block the whole pathway.

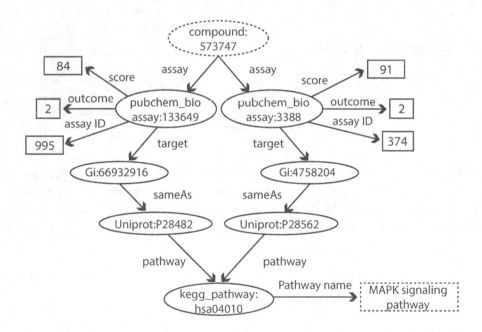

Figure 2.7: Graphical representation of the SPARQL query in the multiple pathway inhibitor case. Note: PubChem compounds (e.g., CID 573747) are active in bioassays which are associated with protein targets or genes (via UniProt), and those protein targes or genes are part of the MAPK, signaling pathway (via KEGG). We identify compounds that have multiple paths, and interact with multiple targets, in this protein.

as a candidate to provide a new direction for the design of inhibitors of both ERK and MKP-1 to reduce cisplatin resistance. After iterating through all the known pathways, we find 36 pathways in which at least two proteins are able to be inhibited simultaneously by at least one compound in PubChem.

The SPARQL query is as follows:

```
PREFIX pubchem: <http://chem2bio2rdf.org/pubchem/resource/>
PREFIX kegg: <http://chem2bio2rdf.org/kegg/resource/>
PREFIX uniprot: <http://chem2bio2rdf.org/uniprot/resource/>
SELECT ?compound_cid (count(?compound_cid) as ?active_assays)

FROM <http://chem2bio2rdf.org/pubchem>
FROM <http://chem2bio2rdf.org/kegg>
FROM <http://chem2bio2rdf.org/uniprot>
WHERE {
?bioassay pubchem:CID ?compound_cid .
?bioassay pubchem:outcome ?activity . FILTER (?activity=2) .
?bioassay pubchem:Score ?score . FILTER (?score>50) .
?bioassay pubchem:gi ?gi .
?uniprot uniprot:gi ?gi .
?pathway kegg:protein ?uniprot .
?pathway kegg:Pathway_name ?pathway_name . FILTER regex(
?pathway_name, "MAPK signaling pathway", "i") .
} GROUP BY ?compound_cid HAVING (count(*)>1)
```

2.3 DISCUSSION

The difficulties of polypharmacology are to explore a combination of targets and then to identify active compounds against these sets of targets. Linking between chemical, biological, systems, and phenotype data is demonstrated to be a promising way to address these problems. For example, linking between bioassay data and marketed drug data enables us to explore the compounds similar to drugs that show polypharmacology. Quinacrine, which has been used as an anthelmintic, and in the treatment of giardiasis and malignant effusions, shows polypharmacology. One compound, loxapine (CID: 71399), is found to show similar polypharmacology with quinacrine. Loxapine is active in both BioAssay 859 and BioAssay 377, whose targets are CHRM1 and ABCB1 respectively. As loxapine tends to be hydrophobic, medicinal chemists would not be surprised that it is active in BioAssay 377, which identifies substrates (or inhibitors) for multidrug resistance transporters. It is also reported that loxapine might get metabolized to amoxapine, a considerably weaker antagonist in BioAssay 859 [Coupet et al., 1985]. Other than loxapine, many identified compounds such as oxybutynin and dexamethasone have been found to show polypharmacology in the literature.

By linking bioassay data to pathways, we can identify the compounds that inhibit at least two of proteins in a pathway, leading to pathway dysfunction. For example, compound CID 6419769 could interact with proteins HSD11B1 and AKR1C4, which are in different branches of C21-Steroid hormone metabolism pathways. The blocking of the pathway might be able to partially explain why CID 6419769 has side effects [Andrews et al., 2003]. In protein-protein interaction networks, two proteins are connected if both physically interact. In terms of polypharmacology, the deletion of one protein does not affect the whole network; but if two connected nodes with high degree are deleted, the network would be disturbed. For example, by linking bioassay to PPI, we found that two compounds (i.e., CID: 460747 and CID: 9549688) are active against two high degree proteins (i.e., PLK1 and TP53) that are associated with cancer.

We note that there are parallel contributions from different data sources and vendors, for example, KEGG and Reactome both provide pathway data. We think that an important part of this work is not just the integration of heterogeneous data, but the integration of sources covering homogeneous kinds of data. Tables 2.2 and 2.3 show the contribution of unique information of homogeneous sources for protein-protein interaction (PPI), and chemogenomics data, respectively. For PPI, HPRD and DIP have 35,645 and 32,976 unique protein pairs, respectively, and the total number of unique pairs in these two datasets is 67,769. Each dataset contributes almost

Table 2.2: Chemogenomics data source distribution

Data source	# of records	Percentage
BindingDB	36,839	12.1%
CTD	95,786	31.5%
DrugBank	10,381	3.4%
Matador	15,843	5.2%
PubChem	146,088	48.1%
QSAR	2,148	0.7%
ALL	303773	

Table 2.3: PPI data source distribution

Datasource	# of records	Percentage
HPRD	35,645	52.6%
DIP	32,976	48.7%
ALL	67,769	

half of the pairs, and there is little overlap between them. The PPI network would not be complete if either dataset was missing. For the chemogenomics data, a chemical protein interaction is recorded as one entry, and all the unique interactions were derived from six datasets. We did not consider another two chemogenomics datasets (i.e., KEGG Ligand and PharmGKB), as KEGG Ligand includes only metabolic molecules rather than chemicals designed for drug discovery, and many drugs in PharmGKB only has names which create problems to link a chemical identifier to a compound. Each dataset only contributes a small number of interactions, so it is not possible represent all chemogenomics data. PubChem BioAssay uses high throughput screening, which allows testing thousands of compounds per experiment, thus yields a large number of chemical protein interactions; but the number of targets studied in PubChem is smaller than that of CTD.

"Pathway" is more complicated than PPI, since each organization could have its own definition of "pathway," which makes integration difficult. For example, a pathway in Reactome is usually composed of a small number of proteins; although the total number of pathways is more than those found in KEGG, the proteins involved in Reactome are far fewer than those in KEGG. We are not able to judge which one is better, so we have to consider all pathway datasets together.

The benefit of integration has ramifications for linking, too. For example, if we take an example of linking a chemical to a pathway via chemogenomics data, a chemical has six directions (i.e., six chemogenomics datasets) to associate with a gene that is mapped to multiple pathways in either KEGG or Reactome. We randomly selected 100 drugs from SIDER and linked each drug to pathways through six chemogenomics datasets. In another four experiments, each experiment only selected one dataset for one domain, instead of using all datasets. If only CTD is selected for chemogenomics data and only KEGG is selected for pathways, the number of paths linking from the 100 drugs to pathways and the number of pathways we found were 6,863 and 178, respectively, compared to 12,240 and 350 when all chemogenomics and pathway data sources were selected (see Table 2.4). CTD and PubChem have the most chemogenomics data; however, if we only used any single dataset, the number of results would decrease rapidly.

Table 2.4: Results of linking sample drugs to pathways

Dataset used	Paths	Gene	Pathways
CTD, KEGG	6,863	763	178
CTD, Reactome	1,157	547	146
PubChem, KEGG	522	33	86
PubChem,Reactome	97	24	18
ALL	12,240	1,181	350

2.4   CONCLUSION

We have created a new Systems Chemical Biology resource called Chem2Bio2RDF, which integrates small molecule, target, gene, pathway, and drug information, and permits cross-source linking with LODD and Bio2RDF. We have demonstrated the utility of this approach in specific examples of polypharmacology and multiple pathway inhibition. We have discussed the importance of integrating, not just heterogeneous data, but also data sources that cover the same kinds of data. However, current data integration is only at the database level, where we have to understand the schema of every database before constructing queries. Hence, the application is quite limited, unless ontology is applied to standardize the schema elements.

CHAPTER 3

Data Representation and Integration Using OWL

3.1 INTRODUCTION

Recent efforts [Belleau et al., 2008, Chen et al., 2010a, Jentzsch et al., 2009] in the Semantic Web have involved conversion of various chemical and biological data sources into semantic formats (e.g., RDF, OWL) and have linked them into large networks. The number of bubbles in Linked Open Data (LOD) has expanded rapidly. This richly linked data allows answering complex scientific questions using SPARQL queries, finding paths among objects [Heim et al., 2010], and ranking associations of different entities [Aleman-Meza et al., 2005, Dong et al., 2010]. Our previous work on Chem2Bio2RDF [Chen et al., 2010a] offers a framework for mining Systems Chemical Biology and chemogenomics data, such as, compound selection in polypharmacology, and multiple pathway inhibitor identification. However, without an ontology and associated annotation, the utility of the resource is semantically limited. For example, results cannot be refined based on criteria regarding the type of relationship between entities (e.g., activation or inhibition between compounds and proteins). Even when it is possible to create a SPARQL query, the lack of ontology increases the complexity of the query. For example, when searching for the targets of a given drug, we have to specify in the SPARQL query exactly which databases we need to search and how to combine the results. SPARQL query construction thus requires understanding of the RDF schema of each data source, greatly increasing query construction complexity. The *owl:sameAs* (or *rdfs:seeAlso*) predicate is used as the primary method for linking multiple data sources sharing common information. Such database-level integration does not satisfy our requirement that a query be constructible in a natural and intuitive manner.

Ontology is a formal description of knowledge as a set of concepts within a domain, and the relationships between those concepts. Web Ontology Language (OWL) is a language for making such descriptions designed for use within the Semantic Web. A variety of ontologies in life sciences have been developed. Gene Ontology (GO) [Ashburner et al., 2000] is arguably the most widely used ontology in life sciences. It aims to formalize the representation of information about biological processes, molecular functions, and cellular components across multiple organisms. As a part of the GO project, the Sequence Ontology consists of a set of terms and relationships used to describe the features and attributes of biological sequences [Eilbeck et al., 2005]. The PRotein Ontology (PRO) describes the relationships of proteins and protein evolutionary families, and represents the multiple protein forms of a gene locus [Natale et al., 2007]. Structurally similar to GO, ChEBI provides ontologies of chemical compounds of biological interest, based on their chemical structural and functional features [Degtyarenko et al., 2008]. Disease Ontology (DO, Schriml et al., 2012) is an ontology for the integration of human disease data. Terms in DO are well-defined using standard references, and linked to well-established, well-adopted terminologies used in other disease presentations, such as MeSH, OMIM, and UMLS. Other domain specific ontologies have also been developed, including pharmacogenomics [Dumontier and Villanueva-Rosales, 2009], ligand protein interaction [Choi et al., 2010, Ivchenko et al., 2011], Disease-Drug Correlation Ontology (DDCO, Qu et al., 2009), biological pathways (BioPAX, Demir et al., 2010), Translational Medicine Ontology [Luciano et al., 2011], and neuromedicine (SWAN, Ciccarese et al., 2008). In particular, several ontologies have been developed recently to formalize chemical biology experiments and provide guidance for data annotation. For example, the Minimum Information About a Bioactive Entity (MIABE, Orchard et al., 2011) provides guidelines for reporting bioactive entities explicitly. The BioAssay Ontology [Visser et al., 2011] standardizes the description of HTS experiments and screening results. DDI [Qi et al., 2010] and OBI [Brinkman et al., 2010] present integrative and semantic frameworks for drug discovery and biomedical investigations, respectively. A number of upper level ontologies, such as the Basic Formal Ontology (BFO), have developed to support domain ontology building. Many of the ontologies are deposited in OBO foundry [Smith et al., 2007] or NCBO bioportal [Noy et al., 2009], for public access.

Using ontologies to integrate data and perform inference is a common practice in life sciences. Baitaluk and Ponomarenko [2010] built IntegromeDB to semantically integrate more than 100 experimental and computational data sources relating to genomics, transcriptomics, genetics, and functional and interaction data concerning gene transcriptional regulation in eukaryotes and prokaryotes. Holford et al. [2010] created logical rules using the Semantic Web Rule Language to answer research questions pertaining to pseudogenes.

Systems Chemical Biology (and its sub-discipline of chemogenomics) is a new discipline, studying how chemicals interact with whole biological systems, the data of which cover a wide range of entities (e.g., compounds, drugs, proteins, genes, diseases, side-effects, pathways, and so on), and various relations between entities (e.g., drug-drug interactions, drug-target interactions, and protein-protein interactions). Within this field, chemogenomics is specifically concerned with ways of modeling the relationships between chemical compounds, genes, and protein targets. Until now, no systematic ontologies have been developed for chemogenomics, or for its parent field of Systems Chemical Biology. In this work, we describe the creation of such an ontology that covers chemogenomics and Systems Chemical Biology, and demonstrate its usage as a knowledge base for various studies.

3.2 SYSTEM AND METHODS

The process for developing Chem2Bio2OWL is shown in Figure 3.1. In particular, our ontology was driven by use cases (available at: http://chem2bio2rdf.org/owl), which are difficult or impossible to address without an ontology and an integrated dataset. Some examples with semantic terms highlighted in bold are the following.

1. What are the *protein targets* of the *drug troglitazone*?

2. Find *PPARG* inhibitors with *molecular weight* less than 500

3. Which *pathways* will be *affected* by *troglitazone*?

4. Find all *bioassays* that *contain* activity data for a particular *target*

5. What *liver-expressed proteins* can *interact* with a given *compound*?

6. Which *proteins* are able to interact with protein *PPARG in vivo*?

7. Which *drugs* are used to *treat diabetes* but have been *withdrawn* from the market?

8. Which *assays* test the *activity* of *troglitazone* against *PPARG* in the *literature*?

CLASSES, RELATIONS, AND DATA PROPERTIES

Once we created an initial set of terms derived from use cases, we defined a set of primary classes: *SmallMolecule, Drug, Protein, Target, Disease, SideEffect, Pathway, BioAssay, Literature*, and *Interaction* (Table 3.1), based partially on the BioPAX classes [Demir et al., 2010]. BioPAX offers a standard, well-defined representation of biological pathway data using OWL, and it has been widely used in biological data integration [Choi et al., 2010, Ruebenacker et al., 2007]. We imported the terms from BioPAX, and made subsequent extensions, based upon our use cases. The primary classes were refined in accordance with current instance data structure. *SmallMolecule, Drug*, and *Protein* were put under *PhysicalEntity*. Their relations with *Disease* and *SideEffect* were elaborated under *Interaction*, which is further classified into *DrugInducedSideEffect, DrugTreatment, DrugDrugInteraction, ProteinProteinInteraction*, and *ChemicalProteinInteraction*. *BioAssay* and *Literature* serve as *Evidence* to support the relations. *Pathway* is treated as a "black box," since its instance data are just pathway names. Other than *Interaction*, we did not intend to further classify other individual major classes.

After major classes were determined, some utility classes were created to help present primary classes, of which a single class is insufficient to present the hierarchical feature. For instance, *ChemicalStructure*, consisting of structure format and structure representation, is considered as a utility class, for presenting the structure of a small molecule. A small molecule may have multiple structure representations; thus, there are several instances of *ChemicalStructure* relating to a small molecule. Without the bearer of small molecule, the instance of *ChemicalStructure* is meaningless.

The relations between entities that associate with properties such as experimental conditions and references were separated as individual classes, and were placed under *Interaction*; otherwise, they were presented as object properties. The Relational Ontology (RO, Smith et al., 2005) was imported to help present basic relations. For example, *ProteinProteinInteraction* not only covers

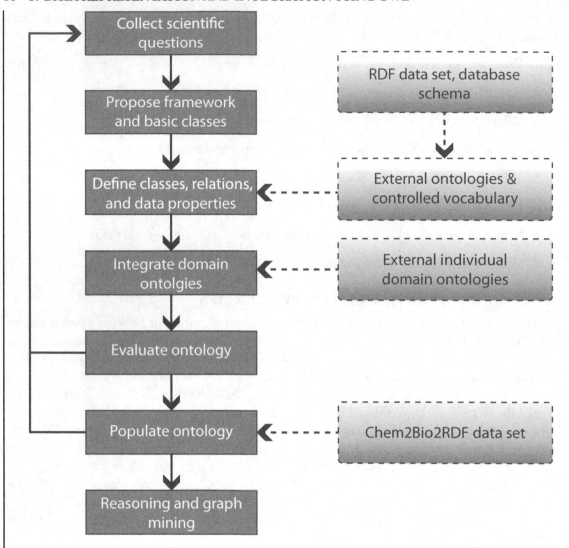

Figure 3.1: Workflow for the development of Chem2Bio2OWL.

the binary relation between two proteins, but also affiliates with its experimental conditions (e.g., organism, interaction type). *Protein* serves as a participant in that interaction. Similarly, *Chemical* and *Protein* serve as participants in *ChemicalProteinInteraction*, which includes other information, such as the strength of interaction. Figure 3.2 shows major classes and their relations.

Data properties appearing in the original database sources were not fully covered. Only the important ones related to our purpose (i.e., chemogenomics and Systems Chemical Biology). This simplified the ontology without losing essential knowledge. The terms including data property name, class name, and relation name were manually mapped to terms in relevant ontologies in OBO and NCBO bioportal, and the terms in the existing ontologies were preferred if multiple terms arose. For example, for a chemical formula, we chose *chemicalFormula*, as this term is used in BioPAX. In addition, the term had to conform to our naming convention. If there were multiple results or no results at all in OBO and NCBO bioportal, we would use the terms from primary databases. A table was created to map data source terms to the standardized ones, and later was applied to annotate instances. The properties of class, object property, and datatype property were further edited in Protégé.

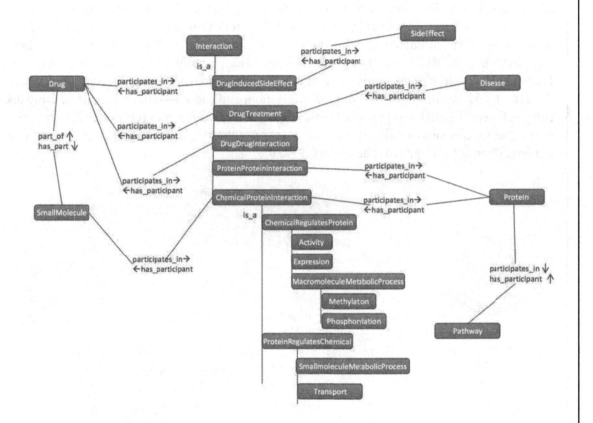

Figure 3.2: Overview of Chem2Bio2OWL. Note: Only parts of classes (presented as nodes) and their relations (presented as edges) are visualized. Some classes in *ChemicalProteinInteraction* are ignored, due to space limit.

CHEMOGENOMIC INTERACTIONS

Classification of chemogenomics interactions (i.e., compound-protein or drug-target) is extremely important and yet complicated [Harland and Gaulton, 2009]. We consider the interaction from two aspects: (1) how chemicals regulate proteins (called *ChemicalRegulatesProtein*) and (2) how proteins regulate chemicals (called *ProteinRegulatesChemical*). ChemicalRegulatesProtein further includes the regulation of protein activity, the expression of protein, and the post-modification of protein. *ProteinRegulatesChemical* includes the catalysis of chemicals and the transportation of chemicals. Interaction types described in the Comparative Toxicogenomics Dataset (CTD) were used as a basis for relational terms, being further developed by the addition of new interaction terms such as *activation* and *inhibition*. The terms were mapped to GO when matching terms were found there. In total, 61 interaction classes were created. The experiment to examine the interaction was presented in BioAssay. The BioAssay outcome includes measurement (e.g., EC50, IC50, Ki, Kd), value, unit (e.g., um, nm), and relation (e.g., $\leq, \geq, =$). Figure 3.3 shows how these terms are used in a small network containing the drug troglitazone, the gene PPAR-Gamma (PPARG), their interaction, and the associated experiment. Only one entry in Chem2Bio2RDF is presented in this figure; there are 43 entries recording this interaction in six chemogenomics databases (available at `http://cheminfov.informatics.indiana.edu/rest/Chem2Bio2RDF/cid_gene/5591:PPARG`).

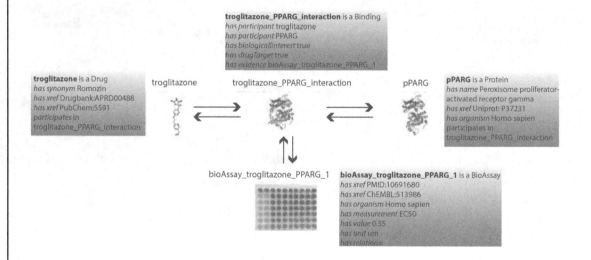

Figure 3.3: Ontological representation of troglitazone, PPARG, and their binding association, tested in a bioassay experiment. Note: The real data are available on the Chem2Bio2RDF website.

3.3 IMPLEMENTATION

Figure 3.4 shows the data integration workflow we used to populate the ontology. Customized Java scripts, along with the OWL API Java package [Horridge and Bechhofer, 2008], were used to automate the annotation of Chem2Bio2RDF data using Chem2Bio2OWL.

Pellet reasoning [Sirin et al., 2007] was then applied to infer new relations. The annotated data plus new relations were uploaded to the Virtuoso triple store for querying. Efforts were made to cope with data redundancy, inconsistency, and provenance. Data redundancy is originated from the homogeneity of data sources of the objects. Chemical compounds, for example, were presented in various formats (e.g., SMILES, InChi, MOL, etc.), and many data sources have their own identifiers for representing compounds. The URI of an individual instance in Chem2Bio2OWL is based on the primary data source ID or a fake ID if the primary ID is unavailable. PubChem, as the largest public compound hub, is considered the primary source for chemicals. Its identifier Compound ID (CID) is used to identify compounds (e.g., `http://chem2bio2rdf.org/chem2b io2owl#compound5591`). Compounds with unknown CIDs were assigned CIDs by searching PubChem using InCHi, a universal structure representation. A fake CID was assigned if the compound was not found in PubChem. Drugs, proteins, and side effects use DrugBank ID, UniProt entry name, and UMLS ID as primary IDs. "Pathway name" is used as the pathway identifier. Diseases can be presented as MESH, OMIM ID, UMLS, or free text, but no universal disease identifier has been agreed upon. Since the Disease Ontology [Schriml et al., 2012] has already mapped its terms to various public disease identifiers, we adopted Disease Ontology ID as the primary ID. The free text in TTD, Diseasome, and other sources were mapped to the disease ontology using string-matching algorithms.

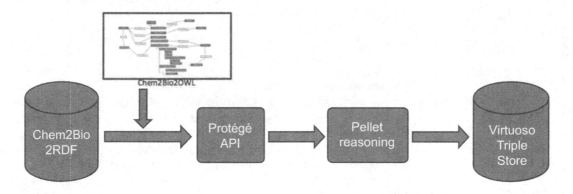

Figure 3.4: Workflow for ontology population.

Maintaining data provenance (i.e., its source and history) is useful for data validation, confidence weighting, and data update and maintenance. The class *UnificationXref* defines a reference to an entity in an external resource that has the same biological identity as the referring entity.

Its data properties *DB* and *ID* present the name of the external source and the related identifier, respectively; *comments* is used to put additional information such as why, who, and how, if needed. For example, compound5591 has ID 5591 in PubChem and ID 9753 in ChEBI; they are represented using class *UnificationXref*. For some assertions (e.g., interactions), *PublicationXref* is applied to record the original paper in which the assertion was published. Table 3.1 shows the statistics of sample instances of primary classes, as well as sample primary data sources. The total number of triples is 3,084,836, and it increases to 4,411,817 after reasoning. These samples were later used for evaluation.

Table 3.1: Primary classes, their descriptions, sample instance data sources, and the number of annotated sample instances [Chen et al., 2010b]

Primary Classes	Description	Sample Instance Data Sources	# of Sample Instances
SmallMolecule	a small bioactive molecule	PubChem, ChEBI	15,509
Drug	a chemical used in the treatment, cure, prevention, or diagnosis of disease	DrugBank, PharmGKB, TTD	6,544
MacroMolecule	a gene or gene expression product	UniProt, HGNC, GOA	12,242
BioAssay	an experiment to measure the effects of some substance on target, cell, or a living organism	PubChem BioAssay, ChEMBL, BindingDB, DPSP	26,861
ChemicalGeneInteraction	genomic response to chemical compounds	ChEMBL, Bind- ingDB, DPSP Ki, TTD, BindingMOAD, DrugBank, CTD, MATA- DOR, ArrayEx- press, KEGG, QSAR	47,282
DrugDrugInteraction	a drug affects the activity of another drug	DrugBank, DCDB	9,690
ProteinProteinInteraction	two or more proteins bind together	HPRD, DIP, Bi- oGrid	54,345
Pathway	a set or series of biological interactions	OMIM, Disea- some, DO	347
Disease	any condition that causes pain, dysfunction, distress, or social problems	OMIM, Disea- some, DO	8,724
SideEffect	undesired effect from a medicine	SIDER	1,385
Literature	a scientific article	Medline	28,392

3.4   USE CASES

In addition to the cases studied before [Chen et al., 2010a] and [Chen et al., 2010b], we applied the annotated data to answer various questions on the Chem2Bio2RDF website (http://ww w.chem2bio2rdf.org). More than 20 SPARQL queries are available at this website. Here we discuss a few examples.

Drug-related target identification. Identification of potential targets for drugs is important for discovering new therapeutic applications, as well as identifying potential undesirable side effects (i.e., "off-target interactions"). These kinds of interactions are described in different ways in various Chem2Bio2RDF datasets: PubChem BioAssay, ChEMBL, and BindingDB containing binding experiments; PharmGKB containing genetic variations upon drug response; CTD and Array Express containing expression data; and KEGG containing interactions in pathways. To answer the question: "*What are the possible targets of a drug (e.g., troglitazone)?*" previously required a complex SPARQL query explicitly referencing each dataset individually [Chen et al., 2010b]. The following SPARQL presents the searching of two chemogenomics databases:

```
PREFIX compound: <http://chem2bio2rdf.org/pubchem/resource/>
PREFIX bindingdb: <http://chem2bio2rdf.org/bindingdb/resource/>
PREFIX drugbank: <http://chem2bio2rdf.org/drugbank/resource/>
PREFIX uniprot: <http://chem2bio2rdf.org/uniprot/resource/>
SELECT ?uniprot_id
FROM <http://chem2bio2rdf.org/pubchem>
FROM <http://chem2bio2rdf.org/drugbank>
FROM <http://chem2bio2rdf.org/bindingdb>
FROM <http://chem2bio2rdf.org/uniprot>
WHERE {

    {?compound compound:CID ?compound_cid . FILTER (?compound_cid=
5591)
    . #Troglitazone PubChem CID is 5591
    ?chemical bindingdb:cid ?compound .
```

The query combines the searching of two databases, BindingDB and DrugBank, which have their own RDF structures. BindingDB and DrugBank use Monomerid and DBID as compound identifiers, respectively, and adopt UniProt and SwissProt ID as target identifiers. They have to be distinct in the SPARQL query. The SPARQL query would become more complicated if other chemogenomics datasets were considered. We can now create a "one-step" query that is independent of the data sources by virtue of our ontology:

```
?target bindingdb:Monomerid ?chemical .
?target bindingdb:ic50_value ?ic50 . FILTER (?ic50<10000) .
?target bindingdb:uniprot?uniprot .
?uniprot uniprot:uniprot ?uniprot_id .}

UNION

{?compound compound:CID ?compound_cid . FILTER (?compound_cid=
5591) .
?drug drugbank:CID ?compound .
? target drugbank:DBID ?drug .
?target drugbank:SwissProtID ?uniprot.
?uniprot uniprot:uniprot ?uniprot_id .}

GROUP BY ?uniprot_id
}
```

The query is interpreted as: A chemical with label "troglitazone" participates in an interaction that is a chemical protein interaction, and the interaction has a participant, which is of type "protein." For troglitazone, other than its primary target PPARG, we found that activities of 10 other targets are associated with the drug, and the gene expressions of 22 targets are either up- or down-regulated by treatment with troglitazone. For example, troglitazone can be metabolized by several cytochrome P450 enzymes (i.e., CYP17A1, CYP2C19, CYP2C8, CYP2C9, and CYP3A4), and also can affect the activity of ABCB11 (i.e., a bile-salt-export pump), which may account for its liver toxicity problems [Funk et al., 2001]. To further explore their interactions, another question might be raised: "*What assays test the activity of troglitazone against PPARG?*" After running the SPARQL below, nine bioassay experiments appeared in five articles were retrieved. Although all assays show the positive activity of troglitazone against PPARG, their values vary under different experiments, the details of which could be further explored via associated references.

```
PREFIX c2b2r: <http://chem2bio2rdf.org/chem2bio2rdf.owl#>
PREFIX bp: <http://www.biopax.org/release/biopax-level3.owl#>
PREFIX ro: <http://www.obofoundry.org/ro/ro.owl#>
PREFIX rdfs: <http://www.w3.org/2000/01/rdf-schema#>
PREFIX rdf: <http://www.w3.org/1999/02/22-rdf-syntax-ns#>

select distinct ?target_name
from <http://chem2bio2rdf.org/owl#>
where  {
?chemical rdfs:label "Troglitazone"^^xsd:string;
ro:participates_in ?interaction .
?interaction rdf:type c2b2r:ChemicalProteinInteraction;
ro:has_participant ?target .
?target rdf:type bp:Protein ;
rdfs:label ?target name .
}
```

The above query can be interpreted as: this interaction has evidence that is a bioassay; the assay has a description and an outcome, and may have a reference, if it exists.

Target inhibitor/activator searching.   Pregnane X receptor (NR1I2) is a transcriptional regulator of the expression of xenobiotic metabolism and transporter genes. It has multiple binding sites, accounting for different functions. Its agonists in the ligand-binding domain would trigger up-regulation of genes, increase the metabolism and excretion of therapeutic agents, and cause drug-drug interactions, but its antagonists counteract such interactions [Ekins et al., 2007]. Due to different binding sites, the two types of compounds may be structurally different. Using Chem2Bio2OWL, we are able to respond to this query " *Find NR1I2 agonists and remove compounds with weight ≥ 500.*" The following SPARQL query was used to retrieve 37 agonists. Their structures are different, with six antagonists retrieved from another query, indicating the significance of classifying the ligands.

```
{
?interaction bp:evidence ?bioAssay ;
?bioAssay rdf:type c2b2r:BioAssay
c2b2r:description ?bioAssayDescription ;
c2b2r:hasOutcome [ c2b2r:measurement ?measurement ;
c2b2r:relation ?relation ;
c2b2r:value ?value ;
c2b2r:unit ?unit ].
optional {?bioAssay bp:xref [c2b2r:title ?title]}
}
```

This query is interpreted as: This interaction is a receptor agonist activity and has a participant that is a small molecule; the molecule has physical property weight smaller than 500, as well as structure with the "openeye can smiles" format.

Thiazolidinedione side effect study.   Thiazolinediones are a class of insulin-sensitizing drugs, widely used to control diabetes. However, several drugs in the class cause severe side effects, resulting in drug withdrawal (troglitazone) or restriction (rosiglitazone). These drugs have a high degree of chemical similarity, but different side effects. Troglitazone is associated with an idiosyncratic reaction leading to drug-induced hepatitis or other liver toxicities [Cohen, 2006], while rosiglitazone is associated with an increased risk of myocardial infarction [Ajjan and Grant, 2008]. The Systems Chemical Biology approach has been shown to have the potential to explain drug side effects [Xie et al., 2009]. Figure 3.5 illustrates two Systems Chemical Biology approaches to investigate the side effects of troglitazone and rosiglitazone. We hypothesize that their related targets might somehow link to disease-related genes/proteins that might explain their side effects. Identification of drug targets and disease-related genes/proteins has two major steps. Via the SPARQL query for drug-related target identification, troglitazone and rosiglitazone were found to interact with 31 and 48 unique targets, respectively. Two approaches could be used to find disease-related genes/proteins, but the first step would map disease terms into Chem2Bio2OWL disease data. We mapped liver toxicity to hepatobiliary disease in the Disease Ontology, which has subclasses such as hepatitis, cholestasis, and hepatorenal syndrome that could be further linked to disease genes in our system (Figure 3.5). ABCB11 is one of the liver disease related genes, and its activity is affected by troglitazone. ABCB11 is involved in liver bile-acid transportation and metabolism (i.e., from GO terms for ABCB11). It is not surprising that a change in its activity can result in liver diseases. Similarly, we mapped heart attack to heart disease in Disease Ontology that includes heart failure, endocarditis, pericarditis, etc., which are linked to seven disease genes. However, no overlap between disease genes and rosiglitazone-related targets were found; therefore, we then decided to seek disease-related targets. First, drugs causing heart disease were sought, and their related targets were identified, grouped, and ranked by the number of their common drugs. A higher ranking indicates a higher possibility of linking to a side effect. The top ten targets are CYP3A4, CYP2C9, ABCB1, CYP1A2, PTGS2, CASP3, CYP2D6, CYP3A5, CYP2C19, and PPARG.

For example, the top target CYP3A4 is shared by 41 out of 181 heart disease related drugs. Some high-ranked targets like CYP3A4 are also affected by troglitazone; nevertheless, the activity of CYP2D6, shared by 24 heart disease related drugs, is affected only by rosiglitazone. It was not found in the troglitazone-related targets. Further literature search indicates that CYP2D6 plays an important role in cardiovascular disease [Lessard et al., 1999]. Although further experimental evaluation would be preferred, this scenario does demonstrate the use of Chem2Bio2OWL to investigate Systems Chemical Biology problems.

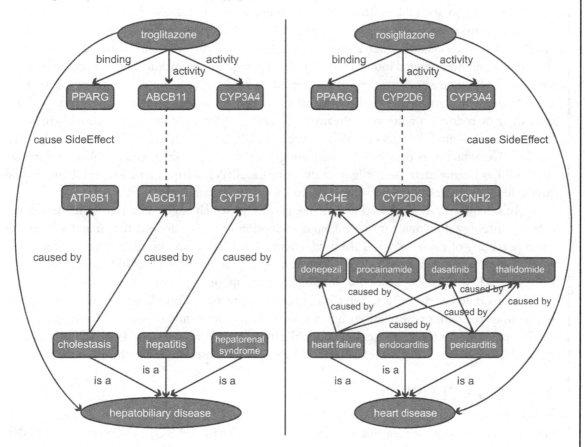

Figure 3.5: Thiazolidinediones side effect study: the left figure shows the association between troglitazone and liver toxicity; the right figure shows the association between rosiglitazone and heart disease.

3.5   DISCUSSION

Many current methods and tools often convert data into RDF triples directly from original databases. They prove useful in their own right to link multiple datasets, but without a formal ontology to model the concepts and their relations, the linked data do not fully demonstrate the capability of semantics, limiting its further usage in data integration and reasoning. In addition, adding evidence and provenance to support assertions is of great importance in keeping track of the record, but has not been well implemented in the current RDFization process. Chem2Bio2OWL provides a high-level structure of the entities and relations according to use cases and instance data structures, and considers evidence and provenance. Besides Chem2Bio2RDF, some public datasets (e.g., Bio2RDF and LODD) can be annotated using Chem2Bio2OWL. Further efforts should be made to align Chem2Bio2OWL with upper level ontologies (i.e., BFO and RO) and other Bio-ontologies. For example, the major classes *PhysicalEntity* and *Disease* are continuant under BFO, and *Pathway*, *Interaction*, and *BioAssay* are occurrent, while other data properties should also be incorporated into BFO so that their usage can be maximized. For example, Utility classes and their subclasses originally served as helper classes for data integration that are actually modeling artifacts. We did not intend to also model major classes, as many of them have their own domain ontologies already (e.g., Disease Ontology for *Disease* class), which can be incorporated into Chem2Bio2OWL accordingly. Since Chem2Bio2OWL was initially based on BioPAX, which was designed for data integration and data exchange of biological pathway data and has been widely used, alignment Chem2Bio2OWL with other basic ontologies would strengthen our collaboration with BioPAX and the OBO community.

In Chapter 1, we discussed knowledge graphs, in which edges have rich semantics; however, in ontology development, we encourage adoption of basic relations that may facilitate the interoperability of the ontology. Take drug-target interaction as an example: A knowledge graph would expect edges annotated by the interaction type, such as binding; but in Chem2Bio2OWL, we separated "interaction" as a class, in which other supporting information about interactions could be modeled as properties. Therefore, in the real practice of building a knowledge graph, it is encouraged to infer new relations based on OWL and other rule languages. Nanopublication might be of help for this purpose. In nanopublication, an assertion states, "drug A binds to target B," and its provenance section refers to the triples in OWL used to derive this new relation.

3.6   CONCLUSION

In summary, we demonstrated how semantic annotation of Systems Chemical Biology data allows meaningful, complex queries to be succinctly specified in SPARQL. We presented an OWL ontology that was used to annotate the Chem2Bio2RDF dataset, and is also available for annotation of other integrative chemogenomics and Systems Chemical Biology datasets. This ontology was developed through a set of use cases, which we believe has made it particularly programmatic. Future work will include aligning this ontology to other widely used ontologies, including the Basic Formal Ontology (BFO) and a variety of other biological ontologies.

CHAPTER 4

Finding Complex Biological Relationships in PubMed Articles using Bio-LDA

There is a vast amount of scientific knowledge embedded in the scholarly literature; indeed, this is the primary vehicle through which such knowledge is transferred. However, the overwhelming number of journal articles published in life sciences each year makes it impossible for scientists to be aware of every paper of interest. This chapter shows that extraction of data on biological entities such as diseases, genes, pathways, drugs, and side effects from recent papers in PubMed, combined with a topic model that organizes these terms and permits calculation of metrics between biological terms and topics, can allow, not just retrieval of relevant papers, but also the identification of potentially interesting relationships between these entities.

4.1 INTRODUCTION

Translational research in medicine is concerned with transforming basic laboratory science into effective patient therapies as quickly as possible. Developing effective treatments requires a cross-discipline understanding of medicine, pharmacology, biology, and chemistry at the physiological, cellular, and molecular levels. The most significant source of knowledge is embedded in published literature. PubMed [Muin et al., 2006] is an online resource that provides over 20 million published articles, and while most are associated with short abstracts, an increasing number is now accompanied by full-text articles. At the same time, sophisticated interdisciplinary research has led to the development and application of powerful methods for generating enormous amounts of new data, which has resulted in an increasingly topical complexity of research articles. This complexity makes it challenging to efficiently discover and evaluate the latest information, trends, and findings deposited in the published literature. For a biomedical researcher, being able to quickly generate and ascertain the significance of associations between chemicals, genes, and diseases is crucial in drug discovery investigations. Thus, generating useful approaches to facilitating knowledge discovery through systematic analysis of abstracts and full-text articles is an important and ongoing challenge.

Natural language processing (NLP) is a common approach to mine biomedical corpora [Cohen and Hunter, 2004, Feldman et al., 2003]. However, NLP largely relies on the syntactic and linguistic structure of documents, and is not by itself able to identify scientific relationships between terms. In contrast, statistical modeling techniques, including Latent Dirichlet Allocation (LDA, Blei et al., 2003), make the automated identification of topics from large document corpora possible [Rosen-Zvi et al., 2004]. LDA, a hierarchical Bayesian model, has been extended to obtain relations between topics and terms [Tang et al., 2008]. Having a specialized and advanced LDA model using life sciences terms may provide an effective way of exploring the biomedical literature.

The LDA model considered in this chapter is a model for a text corpus viewed as a collection of bags of words. It assumes that people write an article with several topics in mind and each topic is associated with a different conditional distribution over a fixed set of words. A collection of documents can be seen as generated by the same set of topics, with a different probability distribution for each document [Hofmann, 1999].

Since the introduction of the LDA model, various extended LDA models have been used in automatic topic extraction from text corpora. Rosen-Zvi et al. [2004] introduced the Author-Topic model that extends LDA to include authorship. Each author is associated with a multinomial distribution over topics. They applied the model to a collection of 1,700 NIPS conference papers and 160,000 CiteSeer abstracts. The primary benefit of the author-topic model is that it allows the explicit inclusion of authors in the document models, providing a general framework for answering queries and making predictions at the level of authors as well as the level of documents. Based on Author-Topic model, Wang and McCallum [2006] presented an Author-Recipient Topic (ART) model for social network analyses, which learns topic distributions based on the direction-sensitive messages sent between entities, adding the key attribute that distribution over topics is conditioned distinctly on both the sender and recipient. Tang et al. [2008] further extended the LDA and Author-Topic model to the Author-Conference-Topic model, which is considered as a unified topic model to simultaneously model the different types of information in an academic network. They found that the proposed method performed well in expertise search and association search. The above studies extended the classic LDA model mainly by incorporating new variables to meet particular demands in the applied area. Other advanced extensions of the LDA model include the supervised Latent Dirichlet Allocation (sLDA, Blei and McAuliffe, 2010).

As for applications of LDA in the biomedical domain, Blei et al. [2006] examined 5,225 free-text items in the Caenorhabditis Genetic Center (CGC) Bibliography using the classic LDA model. They found that like other graphical models for genetic, genomic and other types of biological data, the LDA model estimated from CGC items had better predictive performance than two standard models (i.e., unigram and a mixture of unigrams) both trained with the same data. Zheng et al. [2006] applied the classic LDA model to a corpus of protein-related MEDLINE titles and abstracts and extracted 300 major topics. They found that those topics were semantically coherent with most represented biological objects or concepts. They further mapped those topics to a controlled vocabulary of the Gene Ontology (GO) terms, based on common information. They concluded that the identified topics provided a parsimonious and semantically rich representation of the corpus in a semantic space with reduced dimensionality, and could be used to index text. Mörchen et al. [2008] presented a Topic-Concept model, which extends the basic LDA framework to reflect the generative process of indexing a PubMed abstract with terminological concepts from an ontology. The Topic-Concept model extends the LDA framework by simultaneously modeling the generative process of document generation and the process of document indexing. For each of the concepts in the document, a topic is uniformly drawn, based on the topic assignments for each word in the document, and each concept is sampled from a multinomial distribution over concepts specific to the sampled topic. They applied the model to a large-scale collection of text from PubMed, and found that a number of important tasks for biomedical knowledge discovery could be solved with this Topic-Concept model.

While previous applications of LDA in the biomedical domain had yielded several benefits, few considered the extension of the LDA model to include bio-terms (e.g., gene, compound, disease, etc.) as input parameters. In life sciences, bio-terms are important in determining the literature topics. In this chapter, a Bio-LDA model is proposed, which extends the LDA model by incorporating bio-terms as input variables in the classic LDA model. The associations of the bio-terms are measured based on their topic distributions. This approach is useful for establishing hidden relations between biomedical concepts from literature, in contrast with the commonly used co-occurrence based methods. The identified bio-term associations are evaluated using the known relationships in Chem2Bio2RDF.

Contributions of this chapter include: (1) integration of PubMed and Chem2Bio2RDF using selected bio-terms; (2) development of Bio-LDA, a novel advanced LDA model to mine the latent semantics among topics and biological terms; and (3) measurement of biological term associations based on latent topic distribution.

4.2 MATERIALS AND METHODS

4.2.1 DATABASES

PubMed offers a Web-based and programmatic search service for its content. However, this interface is limited to small- to medium-scale queries, and applying text mining approaches using this interface is not possible. MEDLINE is the primary component of PubMed, where approximately 5,400 biomedical journals are published in the United States and worldwide, and abstracts from 1949 to the present are included. The entire content of MEDLINE is available as a set of text files formatted in XML (i.e., eXtensible Markup Language). In this chapter, the 2010 MEDLINE/PubMed baseline database is used as the primary data source; it contains 617 files and 18,502,916 records.

In order to support information extraction and text mining, a system is developed to load MEDLINE XML files to a relational database, extract bio-terms from MEDLINE, and convert the relational database to RDF triples, as shown in Figure 4.1.

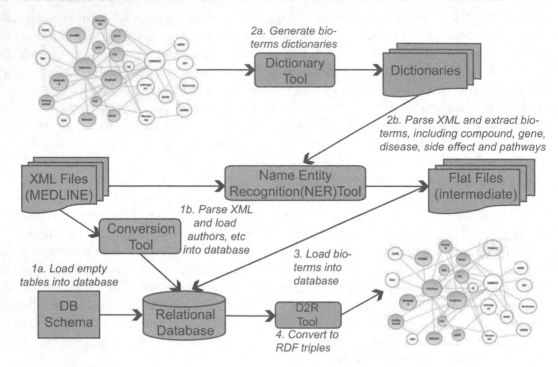

Figure 4.1: Data preprocessing.

The relational database schema used in this system is designed based on the category of bio-terms defined in Chem2Bio2RDF (e.g., compound, drug, gene, disease, side effect, and pathway) and DTD (i.e., Document Type Definition) provided by the National Library of Medicine (NLM). Considering the size of the MEDLINE/PubMed database, a multi-table structure is used in the database to minimize overall lookup times. This MEDLINE database contains several tables: *medline_citation* for title and abstract information; *medline_biblio* for bibliography information; *medline_author* for author information; and *medline_comp* for the compounds mentioned in the citation. All tables contain a PubMed identifier (PMID) in one column, which is usually the foreign key to connect tables, and also the key attribute in the RDF conversion. The Entity-Relation diagram of the local MEDLINE database is shown in Figure 4.2

Information such as citations, authors, journals, and MeSH terms are directly parsed from XML files and loaded into the database. The bio-term information needs to be converted to intermediate files using an extraction method and then loaded into the database. A dictionary extraction method is implemented in this system. Bio-term dictionaries are generated from the following data sources listed in Chem2Bio2RDF: the compound dictionary is generated from PubChem Synonym with the PubChem Compound identifier (CID); the drug dictionary is gen-

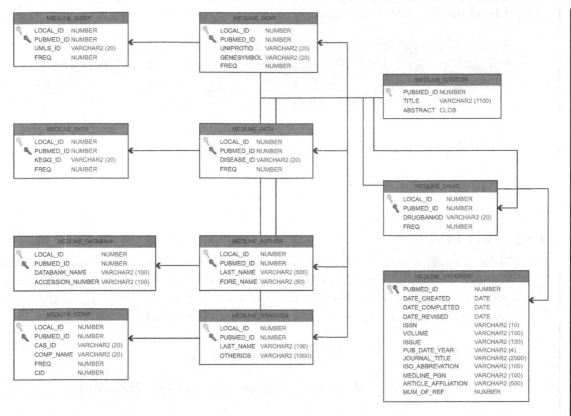

Figure 4.2: Entity-Relation diagram for MEDLINE/PubMed.

erated from DrugBank, and uses DBID as the identifier; the gene dictionary is generated from HGNC, and uses UniProtID as the identifier; the disease dictionary is generated from CTD and uses MeshID as the identifier; the side effect dictionary is generated from SIDER and uses UMLSID as the identifier; and the pathway dictionary is generated from KEGG and uses KeggID as the identifier. The extraction tools parse the XML file and extract the terms based on the pre-generated dictionaries, then save the results to intermediate flat files and load them into the database. Table 4.1 summarizes the dictionary attributes.

The D2R tool is used to convert the MEDLINE/PubMed relational database to Chem2Bio2RDF, which supports data visualization and complex query retrievals. The key attribute for MEDLINE/PubMed triples is PMID. The extracted bio-terms, using the well-known identifiers, are the key to bridging MEDLINE/PubMed with other data sources (i.e., PubChem, UniProt).

Table 4.1: Statistics of the bio-term extraction

Bio-Terms	# of Unique Terms	# of Term-Citation Pairs	# of Unique Citations
Compound	56,383	11,775,891	5,856,084
Drug	2,820	5,624,529	3,427,067
Gene	13,022	5,252,844	3,735,517
Disease	3,848	12,612,636	7,066,084
Side Effect	1,363	10,489,676	6,310,741
Pathway	180	916,754	838,090

4.2.2   BIO-LDA

The Bio-LDA model extends the ACT model and emphasizes bio-terms occurring in the literature. The basic assumption of the Bio-LDA model is that bio-terms of a paper would determine the paper's topics, and each topic then generates the words and determines the publication venue (i.e., a journal). The generative process can be summarized in Figure 4.3.

1. For each bio-term $x = 1, \ldots, B$, draw $\theta_x \sim \text{Dirichlet}(\alpha)$

   For each topic $z = 1, \ldots, T$, draw $\phi_z \sim \text{Dirichlet}(\beta)$, and $\psi_z \sim \text{Dirichlet}(\mu)$

2. For each document $d = 1, \ldots, D$

   Given the vector of bio-terms b_d
   For each word w_i in document d:
   Draw a bio-term $x_{di} \sim \text{Uniform}(b_d)$
   Draw a topic $z_{di} \sim \text{Dirichlet}(\theta_{x_{di}})$
   Draw a word $w_{di} \sim \text{Dirichlet}(\phi_{z_{di}})$
   Draw a journal $j_{di} \sim \text{Dirichlet}(\psi_{z_{di}})$

In this model, the number of possible topics T is fixed. Three continuous random variables, θ, ϕ, and ψ, are involved. For a given set of documents, D^{train}, the aim is to estimate the posterior distribution of those continuous random variables. The inference scheme is based on the observation that

$$P(\theta, \phi, \psi \mid D^{train}, \alpha, \beta, \mu) = \sum_{z,x} P(\theta, \phi, \psi \mid z, x, D^{train}, \alpha, \beta, \mu) P(z, x \mid D^{train}, \alpha, \beta, \mu). \quad (4.1)$$

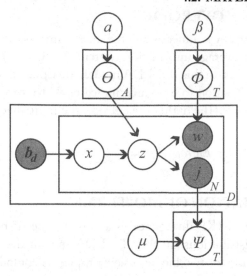

Figure 4.3: Graphical representation of the Bio-LDA model.

In the training process, an empirical sample-based estimation of $P(z, x | D^{train}, \alpha, \beta, \mu)$ is first obtained, using Gibbs sampling.

$$P(z_{di}, x_{di} | z_{-di}, x_{-di}, w, j, \alpha, \beta, \mu) \propto \frac{m_{x_{di}z_{di}}^{-di} + \alpha_{z_{di}}}{\sum_z (m_{x_{di}z}^{-di} + \alpha_z)} \frac{n_{z_{di}w_{di}}^{-di} + \beta_{w_{di}}}{\sum_{w_v} (n_{z_{di}w_v}^{-di} + \beta_{w_v})} \frac{n_{z_{di}j_d}^{-d} + \mu_{j_d}}{\sum_j (n_{z_{di}j}^{-d} + \mu_j)},$$
(4.2)

where the superscript *-di* denotes a quantity, excluding the current instance (e.g., the *di*-th word token in the *d*-th paper). After Gibbs sampling, the probability of a word given a topic ϕ, the probability of a journal given a topic ψ, and the probability of a bio-term given a topic θ, can be estimated as follows:

$$\phi_{zw_{di}} = \frac{n_{z_{di}w_{di}}^{-di} + \beta_{w_{di}}}{\sum_{w_v} (n_{z_{di}w_v}^{-di} + \beta_{w_v})}$$
(4.3)

$$\psi_{zj_d} = \frac{n_{z_{di}j_d}^{-d} + \mu_{j_d}}{\sum_j (n_{z_{di}j}^{-d} + \mu_j)}$$
(4.4)

$$\theta_{xz} = \frac{m_{xz} + \alpha_z}{\sum_{z'} (m_{xz'} + \alpha_{z'})}.$$
(4.5)

With the estimated continuous random variables, θ, ϕ, and ψ, the information content of bio-terms, as well as the association among bio-terms, can be identified.

BIO-TERM ENTROPY OVEROPICS

In information theory, entropy is a measure of the uncertainty associated with a random variable. It is also a measure of the average information content that is missing when one does not know the value of random variables. In this Bio-LDA model, the value of bio-term entropies over topics can be calculated using Equation (4.6), which indicates that bio-terms tend to address a single topic or cover multiple topics. The higher the entropy, the more diverse the bio-term over topics.

$$Entropy(b_i) = -\sum_{z=1}^{T} \theta_{b_i z} \log \theta_{b_i z}. \tag{4.6}$$

SEMANTIC ASSOCIATION OF BIO-TERMS

Kullback-Leibler divergence (KL divergence) is a non-symmetric measure of the difference between two probability distributions. In this Bio-LDA model, the KL divergence, as the non-symmetric distance measure for two bio-terms over topics, is computed using Equation (4.7).

$$KL(b_i, b_j) = \sum_{z=1}^{T} \theta_{b_i z} \log \frac{\theta_{b_i z}}{\theta_{b_j z}}. \tag{4.7}$$

The symmetric distance measure of two bio-terms over topics is the sum of two non-symmetric distances, as shown in Equation (4.8).

$$sKL(b_i, b_j) = \sum_{z=1}^{T} (\theta_{b_i z} \log \frac{\theta_{b_i z}}{\theta_{b_j z}} + \theta_{b_j z} \log \frac{\theta_{b_j z}}{\theta_{b_i z}}. \tag{4.8}$$

In this study, the direction of associations is not important. Thus, the association score is usually calculated by the symmetric distance, unless users specify use of the non-symmetric distance.

4.3   EXPERIMENTAL RESULTS

4.3.1   ANALYZING THE BIO-LDA MODEL RESULTS

In this experiment, the Bio-LDA model is applied to 336,899 MEDLINE/PubMed abstracts (out of $\sim 330\,\text{M}$) published in 2009, which contain 308,686 words, 13,338 extracted bio-terms, and 4,450 journals. Only drug, gene, and disease are considered in this experiment. The output includes the estimated parameters, θ, ϕ, and ψ, and the top lists of words, bio-terms, and journals for each topic. Various numbers of topics, such as 50, 100 and 200, are explored. No significant improvement has been found with the increase of the number of topics. Thus, the 50-topic model is used to optimize efficiency.

Examples of two topics (out of 50 topics in total) with the top 10 representative words, top 10 associated bio-terms, and top 5 related journals are listed in Table 4.2. In this experiment, the

Table 4.2: Representations for selected topics

Topic 13			Topic 14		
Word		**Prob**	**Word**		**Prob**
Patient		0.0177	patient		0.0231
Transplant		0.0149	liver		0.0129
Platelet		0.0074	hepat		0.0126
Studi		0.0066	diseas		0.008
Group		0.0063	studi		0.007
Donor		0.0058	treatment		0.0063
Factor		0.0056	result		0.0059
Risk		0.0054	group		0.0057
Result		0.0053	hcv		0.0056
Graft		0.0053	associ		0.0052
Bio-Terms	**Type**	**Prob**	**Bio-Terms**	**Type**	**Prob**
Thrombosis	DISEASE	0.0855	Hepatitis C	DISEASE	0.0883
Venous Thromboembolism	DISEASE	0.0449	Colitis	DISEASE	0.0784
Heparin	DRUG	0.0417	Hepatitis B	DISEASE	0.0511
Tacrolimus	DRUG	0.0402	Hepatitis	DISEASE	0.0467
Cyclosporine	DRUG	0.0338	Fibrosis	DISEAS	0.0383
VWF	GENE	0.0335	Fatty Liver	DISEASE	0.0274
Thrombocytopenia	DISEASE	0.0274	Ribavirin	DRUG	0.0258
Mycophenolate mofetil	DRUG	0.0259	Liver Cirrhosis	DISEASE	0.0236
IMPACT	GENE	0.0225	Gastroesophageal Reflux	DISEASE	0.0229
ABO	GENE	0.0223	Irritable Bowel Syndrome	DISEASE	0.0222
Journal		**Prob**	**Journal**		**Prob**
Transplant. Proc.		0.0734	epatology		0.064
Transplantation		0.0721	World J. Gastroenterol.		0.0553
Thromb. Haemost.		0.0431	Am. J. Gastroenterol.		0.0532
Thromb. Res.		0.0428	Gastroenterology		0.0477
Transfusion		0.0412	Liver Int.		0.0394

Bio-LDA model is applied to 336,899 MEDLINE/PubMed abstracts (out of ∼ 330 M) published in 2009, which contain 308,686 words, 13,338 extracted bio-terms, and 4,450 journals. Only drug, gene, and disease are considered in this experiment. The output includes the estimated parameters, θ, ϕ, and ψ, and the top lists of words, bioterms, and journals for each topic. Various numbers of topics, such as 50, 100 and 200, are explored. No significant improvement has been found with the increase of the number of topics. Thus, the 50-topic model is used to optimize efficiency.

Table 4.2. The Bio-LDA model provides an unsupervised method for extracting an interpretable representation from a collection of documents. Topic 13 is related to organ transplant and topic 14 is highly related to liver disease (e.g., hepatitis). This Bio-LDA model uses bio-terms, journal information, and word together to characterize the topic, providing a better representation of topics than the simple LDA model that only provides word representation.

Table 4.3 shows the most associated topics for 3 out of 13,338 possible bio-terms. The first bio-term, tuberculosis, denotes an infectious lung disease caused by various strains of mycobacterium. Topic 21 is the major topic associated with tuberculosis, with a conditional probability of 0.8024, and very low probabilities for all other topics. TNFs, tumor necrosis factors, are almost equally distributed across topics 33 and 38, with probabilities of 0.5203 and 0.4149, respectively. The last bio-term, cholesterol, refers to a waxy steroid metabolite found in cell membranes and transported in the blood plasma of all animals. The topics associated with cholesterol are intuitively reasonable. High cholesterol levels usually cause body weight increase (topic 31), and increase the risk of inflammation (topic 33).

The word representations of topics provide an overview of the published literature. Research trends over time could be discovered by applying the Bio-LDA model to different years individually and comparing results.

Table 4.3: Top topics for the selected bio-terms

BioTerm = Tuberculosis (Disease)		
$P(z\|b)$	Topic	Words
0.8024	21	infect, hiv, patient, vaccin, studi, case, tuberculosi, result, year, risk
0.0841	12	gene, protein, cell, express, strain, infect, pathogen, these, host, respons
0.0594	29	protein, bind, activ, structur, cell, domain, interact, these, membran, site
BioTerm = TNF (Gene)		
$P(z\|b)$	Topic	Words
0.5203	33	cell, express, inflamm, activ, inflammatori, induc, increas, alpha, effect, level
0.4149	38	cell, express, activ, immun, respons, induc, mice, cd4, receptor, these
0.0341	30	cell, activ, effect, induc, rat, studi, increas, oxid, level, express
BioTerm = Cholesterol (Drug)		
$P(z\|b)$	Topic	Words
0.3314	31	weight, obes, studi, associ, risk, women, children, group, bodi, increas
0.2926	33	cell, express, inflamm, activ, inflammatori, induc, increas, alpha, effect, level
0.1072	36	insulin, diabet, patient, glucos, level, studi, type, increas, associ, result

4.3.2   COMPARING THE BIO-LDA AND LDA MODELS

In order to compare the output, the topics generated by the LDA model are mapped to the topics generated by the Bio-LDA model. In order to map the topics from these two models, the top 20 words for all topics in the Bio-LDA model are searched over the top 20 words of each topic in the LDA model. The topic with the highest number of shared words is considered as the mapped topic in the LDA model. As shown the word frequency of the top 20 representative words is computed using 50 topics, based on the Bio-LDA model and the LDA model. As shown in

Table 4.4a, 635 distinct words are used to represent topics for the LDA model and 354 distinct words for the Bio-LDA model. 462 words only appear once in the top 20 topic words for the LDA model and 234 words for the Bio-LDA model. This shows that Bio-LDA tends to use fewer words than LDA to represent topics.

Table 4.4: (a) Frequency word sets of the LDA model and the Bio-LDA model; (b) mappings between the Bio-LDA model and the LDA model

Bin	LDA	Bio-LDA	Bin	BioLDA2 LDA	LDA2Bio LDA
1	462	234	1	17	30
2	100	52	2	2	7
3	32	22	3	2	2
4	19	11	4	2	0
5	8	2	5	0	0
6	4	0	6	0	0
7	2	4	7	1	0
8	1	4	8	1	0
9	0	4			
10	4	2			
>10	3	19			
SUM	635	354	SUM	25	39
	(a)			(b)	

In Table 4.4b, the 50 topics in the Bio-LDA model are mapped to 25 topics in the LDA model. Only 17 topics in the Bio-LDA model can be uniquely mapped to topics in the LDA model. The same strategy is used to map the LDA model to the Bio-LDA model. The reverse mapping yields better performance. The 50 topics in the LDA model can be mapped to 39 topics in the Bio-LDA model with about 30 topics have unique mappings. This indicates that the Bio-LDA model is able to cover more topics defined by the LDA model than the LDA model can cover the topics defined by the Bio-LDA model. Table 4.5 shows three examples of mapping LDA topics to Bio-LDA topics. Topic 30 in the LDA model is mapped to topic 25 in the Bio-LDA model, topic 41 mapped to topic 33, and topic 25 mapped to topic 38. There are 11 common words for each mapping.

The overall words used to represent topics in the LDA and Bio-LDA models are explored using a word-cloud tool, *wordle* (http://www.wordle.net/create). Figure 4.4 shows the word-cloud formed by the top 20 words of all 50 topics for the LDA and Bio-LDA models. Some words, such as patient, studi, result, and effect, are shared by both models. However, fewer words are used by Bio-LDA than are used by LDA, and they have different focuses. It also shows that this word-representation of topics is not efficient, and that several words are not meaningful in terms of topic representation; for example, method, group, study, and so on.

Table 4.5: Comparing word representation of topics in the Bio-LDA model to topics in the LDA model

30<-->25		41<-->33		25<-->38	
LDA	Bio-LDA	LDA	Bio-LDA	LDA	Bio-LDA
cell	cell	alpha	cell	cell	cell
induc	Cancer	factor	express	respons	express
apoptosi	express	inflammatori	inflamm	immun	activ
line	Tumor	induc	activ	antibodi	immun
effect	activ	beta	inflammatori	specif	respons
human	Gene	endotheli	induc	antigen	induc
inhibit	protein	increas	increas	anti	mice
death	Induc	inflamm	alpha	gamma	cd4
growth	Human	express	effect	lymphocyt	receptor
prolifer	Growth	activ	level	ifn	these
activ	Inhibit	effect	mice	cd4	cytokin
vitro	studi	tnf	protein	induc	human
p53	Effect	vascular	factor	product	specif
result	Result	growth	studi	activ	regul
increas	Associ	role	tnf	cytokin	function
cycl	line	cytokin	cholesterol	human	antigen
caspas	These	macrophag	result	against	infct
treatment	apoptosi	mmp	role	receptor	role
tumor	Breast	tissu	receptor	cd8	mediat
vivo	Regul	matrix	these	system	signal

Fortunately, the Bio-LDA model can represent topics using not only words but also bio-terms and journals. The word-cloud using the top 20 bio-terms of all 50 topics for the Bio-LDA model is shown in, Figure 4.5, which contains more biologically meaningful words used to represent topics than were found in the LDA model word-cloud.

4.3.3   IDENTIFICATION OF BIO-TERM RELATIONSHIPS WITHIN TOPICS

In the biomedical literature, bio-terms (e.g., drugs, genes, diseases) play an important role in determining topics. The Bio-LDA model makes direct use of bio-terms to improve overall topic generation and word association. As shown in Table 4.6, only 55 words (i.e., 5 drugs, 17 genes, and 33 diseases) are bio-terms, among the 635 unique words. In order to compare the output, the topics generated by the LDA model are mapped to the topics generated by the Bio-LDA model. In order to map the topics from these two models, the top 20 words for all topics in the Bio-LDA

Figure 4.4: Word-cloud of the top 20 words of all topics. LDA (left) and Bio-LDA (right).

Figure 4.5: Word-cloud of the top 20 bio-terms of all Bio-LDA topics.

model are searched over the top 20 words of each topic in the LDA model. The topic with the highest number of shared words is considered as the mapped topic in the LDA model. As shown the word frequency of the top 20 representative words is computed using 50 topics, based on the Bio-LDA model and the LDA model. As shown in Table 4.4a, 635 distinct words are used to represent topics for the LDA model and 354 distinct words for the Bio-LDA model. 462 words only appear once in the top 20 topic words for the LDA model and 234 words for the Bio-LDA model. This shows that Bio-LDA tends to use fewer words than LDA to represent topics.

Table 4.4(a) was generated from the top 20 words of the 50 topics in the LDA model. There are 66 bio-terms (i.e., 9 drugs, 17 genes, and 40 diseases) among the 354 unique words in the Bio-LDA model. Thus, a significant number of bio-terms can be identified in the Bio-LDA model, considering that the Bio-LDA model identifies top bio-terms directly. As shown below,

663 distinct bio-terms, including 145 drugs, 150 genes, and 368 diseases, are identified by the Bio-LDA model.

Table 4.6: Bio-terms associated with topics

Top 20	LDA	Bio-LDA	
	Words	Words	Bio-Terms
Drug	5	9	145
Gene	17	17	150
Disease	33	40	368
bio-terms	55	66	663

There is a general assumption that bio-terms may be associated with each other, if they are in the top-term list of the same topic. Figure 4.6 illustrates an association network drawn from the top five terms from six random selected topics. The connections between two bio-terms are built from various topics using different colors. A solid line indicates that the generated relationships are confirmed by Chem2Bio2RDF (i.e., that there is a relationship in one of the Chem2Bio2RDF datasets that confirms the generated relationship). A dashed line indicates that the generated relationships have not been found in Chem2Bio2RDF, likely indicating very recent findings that are not yet encoded in databases, or associations that are not direct enough to be in a dataset. For instance, the Bio-LDA model suggests that there might be protein-protein interactions between CCND1 and EGFR, since both of them are important targets in the tumor-related diseases, carcinoma and melanoma. Moreover, it suggests that EGFR might also be a target for melanoma, although it was not mentioned in Chem2Bio2RDF. Recently, Chu and Chen [2008] suggests that defective apoptosis in human cancers often results from over-expression or inhibition of BCL2 proteins and proteins with gain-of-function interactions with BCL2, including CCND1 and EGFR.

4.3.4   DISCOVERY OF BIO-TERM ASSOCIATIONS

In traditional text mining, bio-term association is usually calculated based on the co-occurrence of the terms in the literature [Li et al., 2008]:

$$\Theta(b_i, b_j) = \ln(df(b_i, b_j) * N + \lambda) - \ln(df(b_i) * df(b_j) + \lambda). \tag{4.9}$$

Here, $df(b_i)$ and $df(b_j)$ are the number of documents in which bio-terms b_i and b_j are mentioned, respectively; $df(b_i, b_j)$ is the total number of documents in which both bio-terms are mentioned in the same document. N is the size of the document collection. λ is a small constant (e.g., $\lambda = 1$ here) introduced to avoid out-of-bound errors if $df(b_i, b_j)$, $df(b_i)$, or $df(b_j)$ values are 0. $\Theta(b_i, b_j)$ represents the connections between two bio-terms. It is positive when the potential

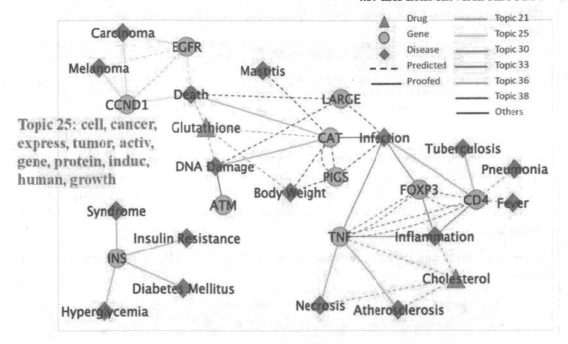

Figure 4.6: An illustration of bio-relationships generated from selected topics.

pairs are over-represented and negative when the pairs are under-represented. The higher the value of $\Theta(b_i, b_j)$, the more significantly the two bio-terms are connected.

However, a limitation of this method is that it cannot detect association between two bio-terms if they are not involved in the same document. For example, both HTR1A and HTR2A do not appear in the same abstracts as venlafaxine based on the PubMed collection. So the calculated association scores are negative, as shown in Table 4.7, which means there shouldn't be any association between venlafaxine and HTR1A or HTR2A. However, venlafaxine is used in the treatment of mental disorders, for example, depressive disorder and anxiety disorder. HTR1A and HTR2A have also been studied in relation to mental disorders. So, in reality, there must be some association between venlafaxine, and HTR1A and HTR2A.

To cover the drawbacks of this co-occurrence based method, a better association approach based on the Bio-LDA topic model is used. In the Bio-LDA model, venlafaxine, HTR1A, and HTR2A are all assigned to topic 10, which is focused on research of mental diseases (e.g., top five word representation for this topic includes *patient*, *studi*, *depress*, *schizophrenia*, and *treatment*). The association score calculated by KL divergence between venlafaxine and HTR1A is quite small, indicating a very strong association between them. It is, therefore, in agreement with the previous explanation.

Table 4.7: Calculated association score for Venlafaxine and HTR1A, HTR2A

Bio-Terms	Co-Occurrence	Bio-LDA
Venlafaxine ~ HTR1A	-11.76	0.34
Venlafaxine ~ HTR2A	-12.72	4.0

In order to get a quantitative measurement of the value of the Bio-LDA model in discovering bio-term associations, the bio-term pairs in Chem2Bio2RDF are used as the standard. As shown in Table 4.8, only a few bio-term pairs are identified using the co-occurrence method. The KL-divergence method, based on the Bio-LDA model, can identify a much larger number of association pairs. The cut-off for the co-occurrence method is 0, and the cut-off for the Bio-LDA model is 5.

Table 4.8: Comparing the co-occurrence method with the Bio-LDA in identifying associated bio-terms

Bio-Terms	Chem2Bio2RDF	Co-Occurrence	Bio-LDA
Disease ~ Gene	412117	266	14,895
Disease ~ Drug	1490	20	228
Gene ~ Drug	5047	28	355
Gene ~ Gene	7593	13	1,282

4.4   APPLICATION TOOLS

As mentioned before, Bio-LDA can identify hidden bio-term associations using the KL divergence score. Two tools are built based on the output of the literature mining method and the Bio-LDA method to discover the literature associations for given pairs of bio-terms.

4.4.1   LITERATURE ASSOCIATION SCORE CALCULATOR (LASC)

In drug discovery, scientists are usually interested in finding out if two items have any correlations, or if any literature mentions are associated with them. Questions like "*Is there any literature relevant to adverse reactions to my drug?*" and "*Which adverse reaction is applicable to my drug?*" are frequently asked. The solution for such questions is to search each drug adverse reaction pair to see if there exists any related literature related, which is inconvenient.

The Literature Association Score Calculator (LASC) is designed to fill this gap (`http://cheminfov.informatics.indiana.edu/huiwang/computeAssociation/LASC.html`); it can search the literature for a set of relationship pairs at the same time. Figure 4.7 shows the interface for the Literature Association Score Calculator. It takes an input file containing the bio-terms in which the user is interested.

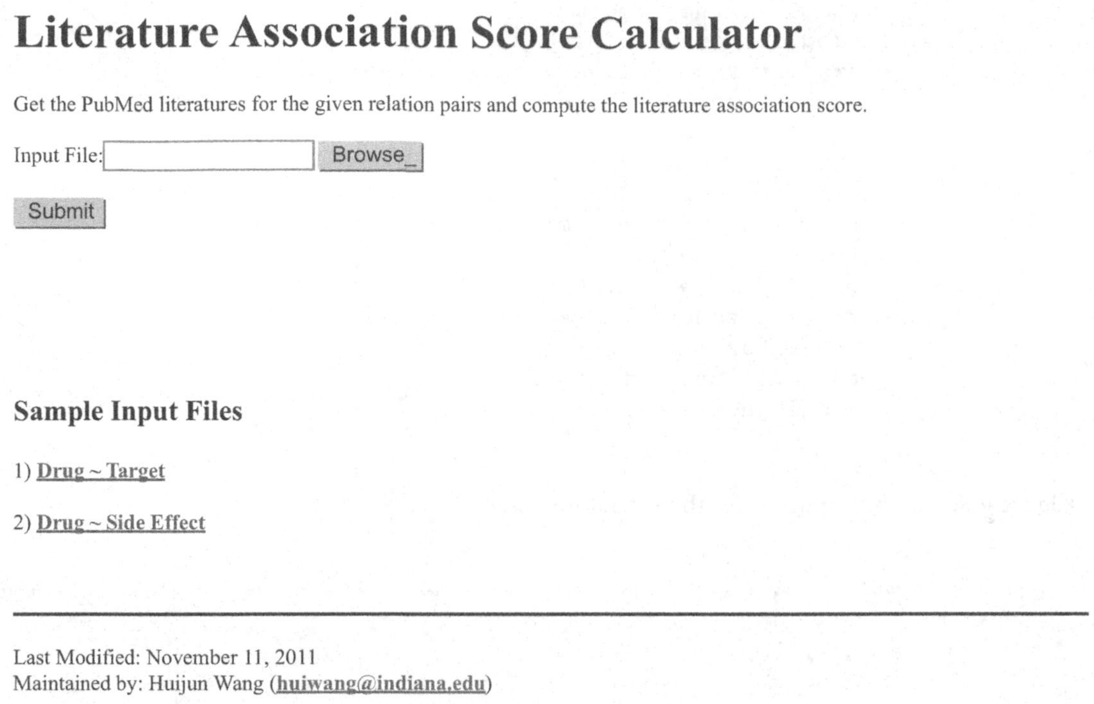

Literature Association Score Calculator

Get the PubMed literatures for the given relation pairs and compute the literature association score.

Input File: [   ] [Browse_]

[Submit]

Sample Input Files

1) **Drug ~ Target**

2) **Drug ~ Side Effect**

Last Modified: November 11, 2011
Maintained by: Huijun Wang (**huiwang@indiana.edu**)

Figure 4.7: Search interface for Literature Associations Score Calculator.

Figure 4.8 shows an example of the input file. Each pair is stored as one line and separated by tabs.

Figure 4.9 shows the result page for the association scores of troglitazone and a set of adverse reactions. Columns 3-5 indicate the number of publications related to each individual term and to both terms; each entry also hyperlinks to the literature page, with search terms highlighted (Figure 4.10). Column is the score calculated using Equation (4.9), which gives the measurement of how likely it is that those two terms appear in the same publication. A score above zero means a relevant finding. The higher the score, the more likely that the two terms occur in the same publication. Columns 7-8 contain the entropy score for each term using the Bio-LDA model and Equation (4.6). Column 9 is the KL divergence score, which represents how closely the two terms are related. The lower the score, the more closely the two terms are related.

```
troglitazone    hepatitis
troglitazone    weight gain
troglitazone    edema
troglitazone    nausea
troglitazone    liver function tests abnormal
troglitazone    jaundice
troglitazone    vomiting
troglitazone    anorexia
troglitazone    fever
troglitazone    abdominal pain
troglitazone    malaise
troglitazone    syncope
troglitazone    congestive heart failure
troglitazone    anemia
troglitazone    fatigue
troglitazone    hyperglycemia
```

Figure 4.8: Sample input file for the Literature Association Score Calculator.

Item1	Item2	Item1 literatures	Item2 literatures	# of co-literatures	co-occur score	Item1 entropy	Item2 entropy	KL divergence
troglitazone	hepatitis	1667	61798	18	1.173	2.242	0.997	16.704
troglitazone	weight gain	1667	28131	38	2.708	2.242	2.328	10.338
troglitazone	edema	1667	44481	19	1.556	2.242	2.911	6.714
troglitazone	nausea	1667	29627	4	0.405	2.242	0.68	18.708
troglitazone	liver function tests abnormal	1667	1	0	-7.419	50	50	50
troglitazone	jaundice	1667	21049	5	0.969	2.242	1.737	15.813
troglitazone	vomiting	1667	34080	2	-0.429	2.242	3.226	15.104
troglitazone	anorexia	1667	10481	1	0.057	2.242	2.526	14.313
troglitazone	fever	1667	67372	1	-1.803	2.242	2.115	13.554
troglitazone	abdominal pain	1667	25360	0	-17.56	2.242	1.883	18.852
troglitazone	malaise	1667	4010	1	1.018	2.242	3.468	13.9
troglitazone	syncope	1667	8780	0	-16.499	2.242	0.814	19.044
troglitazone	congestive heart failure	1667	26132	10	1.446	2.242	0.886	17.132
troglitazone	anemia	1667	60320	5	-0.083	2.242	1.365	18.697
troglitazone	fatigue	1667	38149	2	-0.541	2.242	1.968	17.447
troglitazone	hyperglycemia	1667	18583	61	3.596	2.242	1.81	4.373
pioglitazone	ECG abnormal	1836	3	0	-8.614	1.93	5.32	7.874
pioglitazone	cerebrovascular accident	1836	2361	2	2.144	1.93	3.296	13.942
pioglitazone	weight gain	1836	28131	97	3.548	1.93	2.328	12.624
pioglitazone	heart disease	1836	76636	21	1.016	1.93	1.238	16.272
pioglitazone	Sinusitis	1836	8192	0	-16.526	1.93	1.356	18.512

Figure 4.9: Result page for the Literature Association Score Calculator.

PMID	Publish Year	Title	Abstract
16523797	2006	trial of pioglitazone for the secondary prevention of cardiovascular events in patients with diabetes mellitus type 2: insufficient evidence	the proactive study was a multicentre, multinational, double-blind, placebo-controlled randomised trial that was intended to show a benefit of pioglitazone in the secondary prevention of cardiovascular disease in patients with diabetes. however, the result for the primary composite endpoint was not significant. the most important secondary endpoint (time to death, myocardial infarction or cerebrovascular accident) did show a significant reduction of 16%, but any potential benefit was outweighed by a major increase in the incidence of hospitalisation for heart failure in the pioglitazone-treated group. moreover, in this secondary prevention trial, there was marked undertreatment with statins while no effect of pioglitazone was observed in those who did receive a statin. finally, no adjustment was made for the poorer glycemic control in the placebo group. based on these data, broadening the indication for pioglitazone in patients with diabetes cannot be recommended.
16644631	2006	improvement of glycemic control, triglycerides, and hdl cholesterol levels with muraglitazar, a dual (alpha/gamma) peroxisome proliferator-activated receptor activator, in patients with type 2 diabetes inadequately controlled with metformin monotherapy: a double-blind, randomized, pioglitazone-comparative study.	objective: we sought to evaluate the effects of muraglitazar, a dual (alpha/gamma) peroxisome proliferator-activated receptor (ppar) activator within the new glitazar class, on hyperglycemia and lipid abnormalities. research design and methods: a double-blind, randomized, controlled trial was performed in 1,159 patients with type 2 diabetes inadequately controlled with metformin. patients received once-daily doses of either 5 mg muraglitazar or 30 mg pioglitazone for a total of 24 weeks in addition to open-label metformin. patients were continued in a double-blind fashion for an additional 26 weeks. results: analyses were conducted at week 24 for hba1c (a1c) and at week 12 for lipid parameters. mean a1c at baseline was 8.12 and 8.13% in muraglitazar and pioglitazone groups, respectively. at week 24, muraglitazar reduced mean a1c to 6.98% (-1.14% from baseline), and pioglitazone reduced mean a1c to 7.28% (-0.85% from baseline; p < 0.0001, muraglitazar vs. pioglitazone). at week 12, muraglitazar and pioglitazone reduced mean plasma triglyceride (-28 vs. -14%), apolipoprotein b (-12 vs. -6%), and non-hdl cholesterol (-6 vs. -1%) and increased hdl cholesterol (19 vs. 14%), respectively (p < 0.0001 vs. pioglitazone for all comparisons). at week 24, weight gain (1.4 and 0.6 kg, respectively) and edema (9.2 and 7.2%, respectively) were observed in the muraglitazar and pioglitazone groups; at week 50, weight gain and edema were 2.5 and 1.5 kg, respectively, and 11.8 and 8.9%, respectively. at week 50, heart failure was reported in seven patients (five with muraglitazar and two with pioglitazone), and seven deaths occurred: three from sudden death, two from cerebrovascular accident, and one from pancreatic cancer in the muraglitazar group and one from perforated duodenal ulcer in the pioglitazone group. conclusions: we found that 5 mg muraglitazar resulted in greater improvements in a1c and lipid parameters than a submaximal dose of 30 mg pioglitazone when added to metformin. weight gain and edema were more common when muraglitazar was compared with a submaximal dose of pioglitazone.

Figure 4.10: Publications for a selected pair of items.

4.4.2 ASSOCIATED BIO-TERMS FINDER (ABTF)

The previously described tool, the Literature Association Score Calculator (LASC), calculates the literature association for a set of bio-term pairs. Users need to specify the terms in which they are interested. The Associated Bio-Term Finder (ABTF) tool, (`http://cheminfov.informat ics.indiana.edu/huiwang/computeAssociation/ABTF.html`), on the other hand, does not require the user to specify the term pairs. The search interface and a sample result are shown in Figure 4.11. It searches the whole literature collection to find the relevant bio-terms for a given bio-term. It finds the drug, gene, disease, and a pathway associated with a search query, and returns the frequency of co-occurrence in publications, the co-occurrence score based on the literature model, and the association score based on the Bio-LDA model. The results are ranked based on the Bio-LDA association score (e.g., KL divergence score). The relationships with a KL score of less than 5 are highlighted in yellow. It also provides links to abstracts that contain both terms, similar to Figure 4.10.

Results from this method are compared with results obtained from manual curation (i.e., CTD and Metabase) using diabetes-associated genes. 1,454 genes occur in at least one PubMed abstract together with diabetes, among them, 339 covered by the CTD, and 102 covered by Metabase, and 447 genes covered by both sources. The top 25 rank-ordered results for the Bi-oLDA model, the Language Model (LM), and Term-Frequency (TF), are shown in Table 4.9. The calculated precision for those three models is shown in Table 4.10. It is shown that Bio-LDA is the best measure for this type of association finder, which provides about 80% accuracy for the top 10 returns and 76% for the top 25 returns. Although CTD and Metabase do not record the association of some genes with diabetes found by the Bio-LDA model, research shows that some genes, such as PRKAR1A, are candidates for diabetes [Kwan et al., 2007, Myles et al., 2007].

Table 4.9: Rank-ordered returns for diabetes-related genes

BioLDA			Language Model			Term-Frequency		
GENE	KL-Divergence	Covered	GENE	Score	Covered	GENE	#pubmed	Covered
KCNJ11	1.6	1	B4GALNT1	5.4	0	INS	30908	1
RETN	1.6	1	TAS2R9	5.4	0	LARGE	3852	0
SLC30A8	1.9	1	FOXD4L3	5.4	0	IMPACT	2969	0
SHBG	2.2	0	FOXD4L1	5.4	0	ALB	2334	1
RBP4	2.4	1	ZNF627	4.7	0	GCG	1627	0
G6PC2	2.5	0	PREX1	4.7	1	ACE	957	1
INS	2.8	1	CISD2	4.7	0	CAD	826	0
CDKAL1	2.9	1	UTS2	4.7	1	HR	799	0
TCF7L2	2.9	1	FOXK2	4.7	0	REST	786	0
IGF2BP2	3.1	1	FOXJ3	4.7	0	REN	774	0
PRKAR1A	3.1	0	WFS1	4.5	1	SET	762	0
ENPP1	3.3	1	SLC19A2	4.3	0	TG	736	1
SLC2A2	3.5	1	PTPRN	4.3	0	INSR	635	1
HNF4A	3.7	1	UTS2R	4.3	1	MET	614	0
CNDP1	3.8	1	HPCAL1	4.3	0	TNF	605	1
PPARGC1A	3.8	1	PPP1R2	4.3	0	LEP	538	1
HHEX	3.9	1	CAPN5	4.2	0	HBA1	522	1
GCKR	4.2	0	PRSS16	4.2	0	CRP	450	1
TPO	4.4	0	HBA1	4.1	1	SST	444	0
ADIPOQ	4.4	1	CAPN10	4.1	1	PLG	428	0
ADIPOR1	4.4	1	KCNJ9	4.1	0	CAT	404	1
SDHB	4.5	1	KCNJ11	4.0	1	CD4	384	1
IRS2	4.7	1	ALMS1	4.0	0	ADA	285	1
SLC2A4	4.7	1	IAPP	4.0	0	CA2	284	0
GIP	4.8	0	CACNA1D	4.0	0	PTH	244	0

Table 4.10: Computed precision

	Precision		
Rank	BioLDA	LM	TF
1	1.00	0.00	1.00
2	1.00	0.00	0.50
3	1.00	0.00	0.33
4	0.75	0.00	0.50
5	0.80	0.00	0.40
6	0.67	0.17	0.50
7	0.71	0.14	0.43
8	0.75	0.25	0.38
9	0.78	0.22	0.33
10	0.80	0.20	0.30
11	0.73	0.27	0.27
12	0.75	0.25	0.33
13	0.77	0.23	0.38
14	0.79	0.29	0.36
15	0.80	0.27	0.40
16	0.81	0.25	0.44
17	0.82	0.24	0.47
18	0.78	0.22	0.50
19	0.74	0.26	0.47
20	0.75	0.30	0.45
21	0.76	0.29	0.48
22	0.77	0.32	0.50
23	0.78	0.30	0.52
24	0.79	0.29	0.50
25	0.76	0.28	0.48

The results also confirm that a KL divergence score of less than 5 is a good cutoff for accurate predictions. For the general-purpose research, which may require comprehensive candidate lists, users can work with results with higher KL divergence scores.

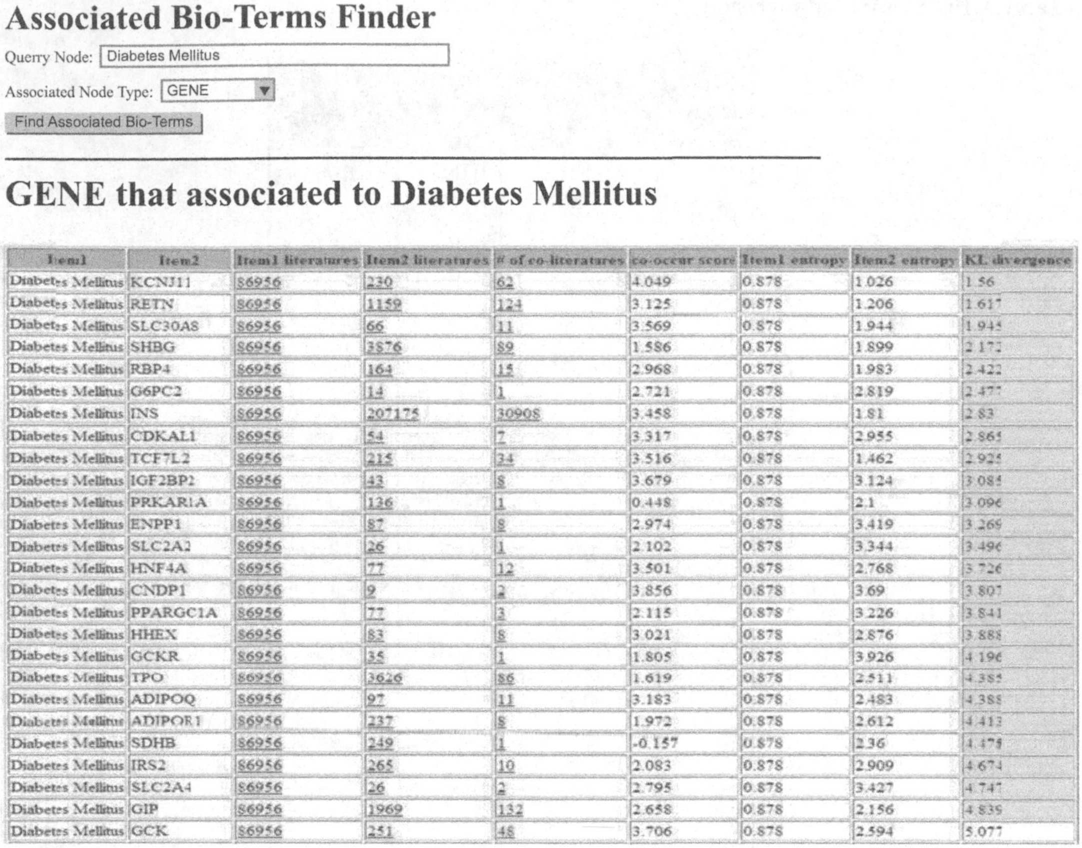

Associated Bio-Terms Finder

Querry Node: [Diabetes Mellitus]

Associated Node Type: [GENE ▼]

[Find Associated Bio-Terms]

GENE that associated to Diabetes Mellitus

Item1	Item2	Item1 literatures	Item2 literatures	# of co-literatures	co-occur score	Item1 entropy	Item2 entropy	KL divergence
Diabetes Mellitus	KCNJ11	86956	230	62	4.049	0.878	1.026	1.56
Diabetes Mellitus	RETN	86956	1159	124	3.125	0.878	1.206	1.617
Diabetes Mellitus	SLC30A8	86956	66	11	3.569	0.878	1.944	1.944
Diabetes Mellitus	SHBG	86956	3876	89	1.586	0.878	1.899	2.173
Diabetes Mellitus	RBP4	86956	164	15	2.968	0.878	1.983	2.422
Diabetes Mellitus	G6PC2	86956	14	1	2.721	0.878	2.819	2.477
Diabetes Mellitus	INS	86956	207175	30908	3.458	0.878	1.81	2.83
Diabetes Mellitus	CDKAL1	86956	54	7	3.317	0.878	2.955	2.865
Diabetes Mellitus	TCF7L2	86956	215	34	3.516	0.878	1.462	2.925
Diabetes Mellitus	IGF2BP2	86956	43	8	3.679	0.878	3.124	3.085
Diabetes Mellitus	PRKAR1A	86956	126	1	0.448	0.878	2.1	3.096
Diabetes Mellitus	ENPP1	86956	87	8	2.974	0.878	3.419	3.269
Diabetes Mellitus	SLC2A2	86956	26	1	2.102	0.878	3.344	3.496
Diabetes Mellitus	HNF4A	86956	77	12	3.501	0.878	2.768	3.726
Diabetes Mellitus	CNDP1	86956	9	2	3.856	0.878	3.69	3.807
Diabetes Mellitus	PPARGC1A	86956	77	3	2.115	0.878	3.226	3.841
Diabetes Mellitus	HHEX	86956	83	8	3.021	0.878	2.876	3.888
Diabetes Mellitus	GCKR	86956	35	1	1.805	0.878	3.926	4.196
Diabetes Mellitus	TPO	86956	3626	86	1.619	0.878	2.511	4.385
Diabetes Mellitus	ADIPOQ	86956	97	11	3.183	0.878	2.483	4.388
Diabetes Mellitus	ADIPOR1	86956	237	8	1.972	0.878	2.612	4.413
Diabetes Mellitus	SDHB	86956	249	1	-0.157	0.878	2.36	4.478
Diabetes Mellitus	IRS2	86956	265	10	2.083	0.878	2.909	4.674
Diabetes Mellitus	SLC2A4	86956	26	2	2.795	0.878	3.427	4.747
Diabetes Mellitus	GIP	86956	1969	132	2.658	0.878	2.156	4.839
Diabetes Mellitus	GCK	86956	251	48	3.706	0.878	2.594	5.077

Figure 4.11: Search interface for Associated Bio-Terms Finder.

4.5 CONCLUSION

This chapter demonstrates that indexing PubMed using bio-terms, and implementing the Bio-LDA model on PubMed, are valuable in mining biomedical literature. The results indicate that Bio-LDA can yield better coverage for finding hidden relations for biological terms than the LDA model and the language model. Based on the proposed algorithms, two application tools, the Literature Association Score Calculator (LASC) and the Associated Bio-Terms Finder (ABTF), are developed, to help users find complex biological relationships in PubMed using the proposed methods.

CHAPTER 5

Integrated Semantic Approach for Systems Chemical Biology Knowledge Discovery

Recent advances in drug discovery require the understanding of how compounds affect the whole system, instead of a single target. There has been an increasing demand for identifying interesting relationships between pairs of biological entities (e.g., compounds, genes, diseases, side effects, and pathways) in large biological systems. A tremendous amount of relevant data has been published through either database collections or publications. Previous work demonstrated an integrated semantic Systems Chemical Biology data source, Chem2Bio2RDF, and a data mining approach, Bio-LDA, to quantitatively measure bio-term associations hidden in the literature. This chapter introduces an integrated semantic approach for Systems Chemical Biology knowledge discovery, which first detects the potential associations and paths in Chem2Bio2RDF, and then uses the Bio-LDA model to provide contextual evaluation of those associations and paths.

5.1 INTRODUCTION

In the last few decades, drug discovery research on molecules has focused on finding chemicals that bind to only one target (i.e., the specified disease target). However, induction of a disease state is often the result of an incredibly complex combination of molecular events. As a result, many promising drug candidates fail in the last and the most expensive clinical phase, because of their off-target effects. For instance, Pfizer discontinued torcetrapib, a drug-targeting Cholesteryl Ester Transfer Protein (CETP) Inhibitor for treating cardiovascular diseases, due to off-target effects. Research of this drug cost Pfizer $800 million before its termination (`http://www.gene ngnews.com/keywordsandtools/print/3/21572/`).

Recent advances in drug discovery use network and systems approaches, which take protein targets back to their physiological context and study the complex biological responses caused by molecules. If successful, this approach can significantly reduce the time and cost of new drug development [Butcher, 2005].

A chemical and biological interaction network is a prerequisite for the systematic approach mentioned above. Previous chapters show that Semantic Web technologies can provide machine-understandable semantics for integrating heterogeneous data sources and retrieving information using SPARQL queries. Chem2Bio2RDF has been developed to link compounds, genes, diseases, side effects, and pathways, to form an interaction network.

Understanding complex relationships in a large-scale network is a critical step towards knowledge discovery in drug research. For instance, chemogenomics studies how compounds affect a single gene target; polypharmacology considers compounds' multi-target effects; and Systems Chemical Biology explores how compounds interact with the whole system.

Several works have studied semantic network relationships. For example, Anyanwu and Sheth [2002] defined three types of complex relationships, *ρ-Path association*, *ρ-Join association*, *and ρ-Iso association*, to discover complex relationships. However, those methods are not scalable for large graphs, like Chem2Bio2RDF. Some graph mining methods have been applied to large biological networks to study specific interactions, such as compound-target interaction [Yamanishi et al., 2010], disease-gene association [Hwang et al., 2011], and protein-protein interaction. General exploration of non-specific associations is less studied, except for some data mining methods for discovering hidden connections between drugs, genes, and diseases [Frijters et al., 2010].

In this chapter, an integrated approach with both graph mining and data mining is developed for discovering potential relationships. Graph mining methods are applied to the large-scale network, Chem2Bio2RDF, to detect possible associations and paths between biological entities. In addition, the Bio-LDA model, which extracts contextual information on topics of bio-terms and interprets semantic associations, is used to provide contextual evaluation of detected associations and paths.

5.2   DATASETS

This integrated approach takes advantage of both the Chem2Bio2RDF and the Bio-LDA results mined from the PubMed literature. Chem2Bio2RDF covers 25 biomedical datasets. The data is further simplified using the Chem2Bio2OWL ontology mentioned in Chapter 3, which focuses on compounds, genes, diseases, side effects, and pathways, and the interactions among them. On the other hand, the biological terms defined by Chem2Bio2OWL are indexed in the abstracts of PubMed articles. The Bio-LDA model is then applied to those indexed abstracts to yield a latent text association measurement of bio-terms.

5.3   METHODS

As mentioned in Chapter 4, the association of two bio-terms in the literature can be measured by KL-divergence. This association score can be combined with pre-knowledge of bio-terms (i.e., Chem2Bio2RDF) for novel knowledge discovery.

5.3.1   ASSOCIATION PREDICTION

Two bio-terms can be associated if there exists a path between them, or if two bio-terms have similar chemical or biological activities. The graphic definition of three types of semantic association is shown in Figure 5.1 [Anyanwu and Sheth, 2002]. However, the number of association pairs is usually huge for a large network. The association score from the Bio-LDA model can be used to rank and select the most interesting pairs from the candidate pool.

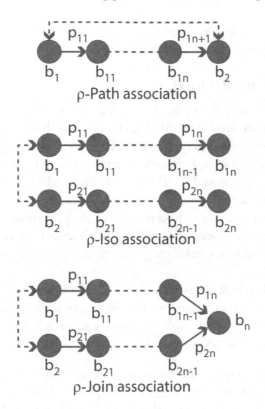

Figure 5.1: Semantic associations [Anyanwu and Sheth, 2002].

In Chem2Bio2RDF, there are eight types of relations: compound-compound, compound-gene, compound-disease, compound-side effect, compound-pathway, gene-gene, gene-disease, and gene-pathway. For a given source, it is relatively easy to find its associated target with a given type in Chem2Bio2RDF. For instance, finding compounds that target genes can be achieved by finding direct relations among compound-gene pairs. However, users usually are not only interested in known linkages, but also want to get information about possible indirect linkages, which may offer opportunities to find hidden relations (e.g., ρ-Path association). The quality of those indirect relations can be evaluated using the calculated association scores from the Bio-LDA topic model. For instance, in order to find possible indirect linkages for a given gene-compound pair, four extended associations in Chem2Bio2Rdf are considered: gene-disease-compound, gene-compound-compound, gene-pathway-compound, and gene-gene-compound. Association scores are computed on those outputs from Chem2Bio2RDF. A valid extended association is defined as follows:

$$\left.\begin{array}{c} sKL(b_1, b_2) \leq c_T \\ Association(b_1, b_2) \end{array}\right\} \Rightarrow \hat{R}_T(b_1, b_2), \tag{5.1}$$

where $Association(b_1, b_2)$ indicates possible semantic ρ-associations from Chem2Bio2RDF, and $sKL(b_1, b_2)$ is the association score calculated using Equation (4.8), described in Chapter 4.

5.3.2 ASSOCIATION SEARCH

In network analysis, the task of association search can be formalized as a task of pathfinding in a graph. In this study, a semantic network (e.g., Chem2Bio2RDF, or a simplified graph, ChemBioSpace) can be represented as a graph $G = (V, E)$, where $v \in V$ represents an entity (e.g., drug or gene) in the network; and $e_{ij}^r \in E$ represents a relationship with property r (e.g., drug-target interaction) between entities v_i and v_j. The relationship can be directional or bi-directional. The goal of association search is to find relationship sequences from v_i to v_j. There is an assumption that no entity will appear in a given association more than one time. Thus, the process of association search from one entity to another is defined as: *Given an association query* (v_i, v_j), *where* v_i *denotes the source entity and* v_j *denotes the target entity, an association search finds possible associations* $\alpha_k(v_i, v_j)$ *from* v_i *to* v_j.

There are two subtasks in an association search: finding possible associations between two bio-terms and ranking the associations. In this work, the association search problem is formed as a near-shortest association search. A two-stage approach is proposed to find near-shortest associations for an association query (v_i, v_j).

1. Finding the shortest association. This process seeks the shortest associations from all entities $v \in V \setminus v_j$ in the network to the target entity v_j (including the shortest association from v_i to v_j, with length L_{min}). In a graph, the shortest path between two nodes can be found using a heap-based Dijkstra algorithm, to quickly find the shortest association that can achieve a complexity of $O(n \log n)$.

2. Finding the near-shortest associations. Based on the length of a shortest association L_{min} and a pre-defined parameter β, the algorithm requires enumeration of all associations that are less than $(1 + \beta)L_{min}$ by a depth-first search. The length of an association is constrained to be less than a pre-defined threshold. This length constraint can reduce the computational cost.

The obtained associations are ranked according to the accumulated KL divergence scores obtained from the Bio-LDA model.

5.3.3 ASSOCIATION EXPLORATION

Association search finds possible associations between two bio-terms, and ranks paths that connect them. In some cases, user may be interested in how a bio-term is associated with a certain type of bio-terms. For example, what is the largest number of possible pathways associated with cardiovascular disease? The algorithm for the association exploration is as follows.

1. Use a Breadth-First Search (BFS) algorithm to find all possible paths with the given length.

2. Check if the end node type matches the specified node type.

3. Calculate the association scores for the qualified paths and rank them.

Association exploration is different from association prediction, although both methods look for certain types of nodes for a given start node. Association prediction specifies a meta travel path, and is ranked by the calculated association scores of the start and end nodes. Association exploration ignores how the start node is connected to the end node, as long as the connection length is within a certain range. The results are based on the overall association score of the whole path, instead of only the start and end nodes.

5.3.4 CONNECTIVITY-MAP GENERATION

The concept of molecular connectivity maps has gained popularity in Systems Chemical Biology [Lamb et al., 2006]. It helps researchers study and compare the molecular therapeutic/toxicology profiles of many candidate drugs. In this chapter, a computational approach is proposed to build interest-specific connectivity maps, for example, to build disease-specific gene-drug connectivity maps, based on both genomic data sources and the literature. The input query for the connectivity map is $((v_i, t), t_1, t_2)$, where (v_i, t) is the specified interest and its type (i.e., Alzheimer's Disease, Disease), and t_1 and t_2 are the bio-term types that form the connectivity maps (i.e., drug and gene). In this study, the candidate bio-terms in t_1 are identified and refined based on the genomic data sources (i.e., Chem2Bio2RDF), and the candidate bio-terms in t_2 are the bio-items that can interact with the candidate bio-terms in t_1 based on Chem2Bio2RDF linkages. The association score calculated on the Bio-LDA model yields their connection scores. Here, the disease-specific gene-drug connectivity map is used as an example to show the process.

1. Specify a disease.

2. Identify genes that are related to the given disease from prior knowledge (i.e., Chem2Bio2RDF).

3. Expand the genes based on gene-gene interactions. This step can be ignored if the user does not want to count gene-gene interactions.

4. Combine and re-rank genes identified in steps 2 and 3.

5. Find drugs that can target one or more genes from the gene set given by step 4. The drugs are ranked based on an accumulated score of the importance of the targeted genes.

6. Calculate the association score based on the Bio-LDA model for the gene set from step 4 and the drug set from set 5 to form the connectivity maps.

5.3.5   Chem2Bio2RDF EXTENSION

All the above methods use Chem2Bio2RDF as the fundamental network to perform the search and evaluate the results using the Bio-LDA results on PubMed. Chem2Bio2RDF usually lacks discovery compounds or virtual compounds. In order to apply the integrated methods mentioned above, a sub-graph is created to connect the novel compounds with known terms in Chem2Bio2RDF. The sub-graph is then merged with Chem2Bio2RDF as a new network to support the knowledge discovery process. The sub-graph can be created using the following sources.

1. Experiment measurements (i.e., high throughput screen (HTS)) for compound-target interaction data.

2. Similarity measurements (i.e., structure similarity, using the Tanimoto coefficient) to yield compound-compound interaction data.

3. Model predictions (i.e., predicted compound-target interactions), using known pharmacology data.

5.4   APPLICATION TOOLS

A set of Web tools, an association predictor, an association searcher, and an association explorer, are developed, based on the methods discussed in the above section. The Web version of the Connectivity Map Generator is not available yet; only the local version is in use right now.

5.4.1   ASSOCIATION PREDICTOR

The association predictor is based on the algorithm mentioned in Section 5.3.1 (http://cheminfov.informatics.indiana.edu/huiwang/association/predication.html). The search interface is shown in Figure 5.2, which takes a start node name and type, and an end node type. Users can specify the intermediate node type and the cut-off value for predictions. The tool first searches Chem2Bio2RDF for qualified paths. The association scores of the start nodes and end nodes given by qualified paths are calculated and ranked. In order to produce a broader predication, both the co-occurrence score and KL divergence score are used.

Figure 5.3 shows the predicted drugs for ABL1 through disease. The top table lists five drugs that are directly connected to ABL1 in Chem2Bio2RDF. The left bottom table shows predicted drugs using the Language Model, with co-occurrence score great than 1. The middle bottom table shows predicted results using the Bio-LDA model. The right bottom table is the

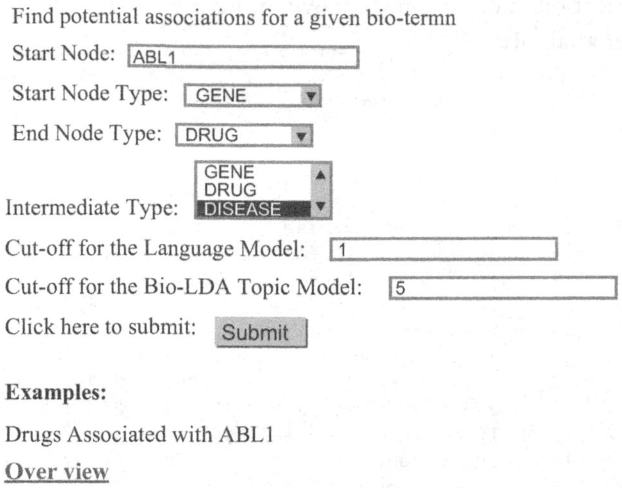

Find potential associations for a given bio-termn

Start Node: [ABL1]

Start Node Type: [GENE ▼]

End Node Type: [DRUG ▼]

Intermediate Type: [GENE / DRUG / DISEASE ▲]

Cut-off for the Language Model: [1]

Cut-off for the Bio-LDA Topic Model: [5]

Click here to submit: [Submit]

Examples:

Drugs Associated with ABL1

Over view

Figure 5.2: Interface for associations predictor.

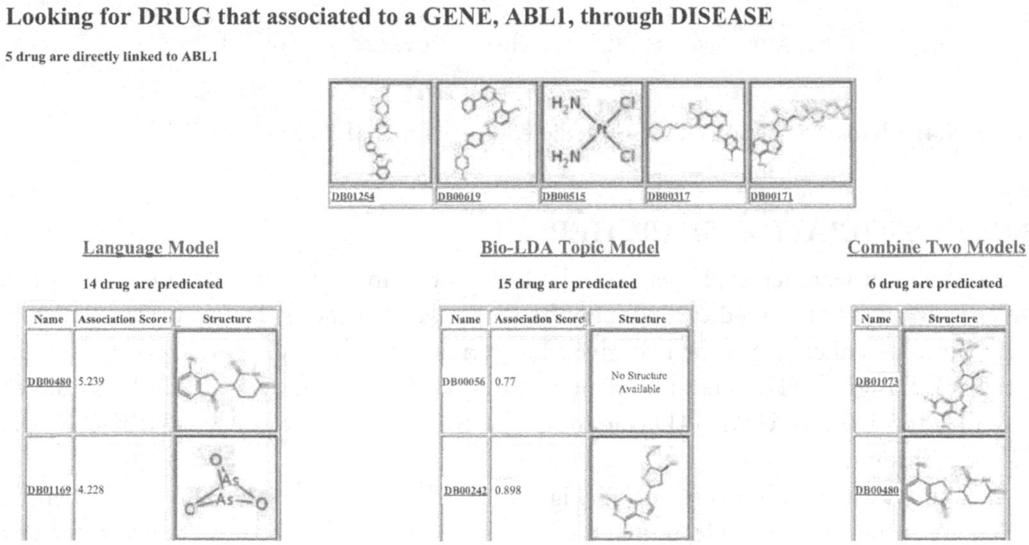

Looking for DRUG that associated to a GENE, ABL1, through DISEASE

5 drug are directly linked to ABL1

DB01254 DB00619 DB00515 DB00317 DB00171

Language Model	Bio-LDA Topic Model	Combine Two Models
14 drug are predicated	15 drug are predicated	6 drug are predicated

Name	Association Score	Structure
DB00480	5.239	
DB01169	4.228	

Name	Association Score	Structure
DB00056	0.77	No Structure Available
DB00242	0.898	

Name	Structure
DB01073	
DB00480	

Figure 5.3: Association prediction of drugs for gene ABL1 through the gene-disease-drug path.

union set of the previous two tables. Each table is sorted based on association scores, and provides both drug structures and linkages to the original Drugbank page. Each model name is a hyperlink to the network connection information, as shown in Figure 5.4, which can be used as a Cytoscape input file for further analysis.

```
P00519   DB00619   gene-drug        1.052   0.97    3.035
P00519   DB01254   gene-drug        1.052   1.377   3.44
P00519   DB00515   gene-drug        1.052   1.714   22.542
P00519   DB00317   gene-drug        1.052   1.129   19.372
P00519   DB00171   gene-drug        1.052   3.866   15.708
P00519   D009369   gene-disease     1.052   1.434   14.798
P00519   D054198   gene-disease     100     100     100
P00519   D015464   gene-disease     100     100     100
P00519   D010051   gene-disease     1.052   1.634   15.446
P00519   D007938   gene-disease     1.052   0.587   0.91
P00519   D007951   gene-disease     1.052   4.901   3.829
P00519   D011471   gene-disease     1.052   2.721   14.428
P00519   D008545   gene-disease     1.052   2.021   21.983
P00519   D015470   gene-disease     100     100     100
P00519   D009190   gene-disease     1.052   0.197   1.289
P00519   D013921   gene-disease     1.052   1.465   8.427
P00519   D020246   gene-disease     1.052   2.393   17.981
P00519   D009101   gene-disease     1.052   0.513   1.967
P00519   D000755   gene-disease     1.052   3.117   12.981
D054198  DB00023   disease-drug     100     100     100
D015464  DB01008   disease-drug     100     100     100
D015470  DB00987   disease-drug     100     100     100
D009101  DB01041   disease-drug     0.513   1.414   2.027
```

Figure 5.4: Cytoscape network file for predicted drugs for ABL1 using Bio-LDA.

5.4.2 ASSOCIATION SEARCHER

The association searcher implements methods described in Section 5.3.2 (http://cheminfov. informatics.indiana.edu:8080/chembiospace/pathfinder/hychembiospace.html). It finds the top ranked paths for two given bio-terms (e.g., drug, gene, disease, side effect, and pathway). DrugBank ID is used to identify drugs, UniProt ID for genes, Mesh ID (mentioned in CTD) for diseases, UMLS ID (mentioned in SIDER) for side effects, and KEGG pathway ID for pathways.

The search interface is shown in Figure 5.5. The inputs include the start and end nodes, which are names or IDs of bio-terms; the length of path allowed to travel between the start and end nodes, defaulting to 3; and the number of top-ranked paths the user wants to see, defaulting to 20. Name autocompletion is implemented to allow users to easily type search terms.

The search results show the top-ranked connections between the start and end nodes using Adobe Flash. Users can require the results in a text format, by providing their email address. The output of the full results will be sent to the email address, in two parts.

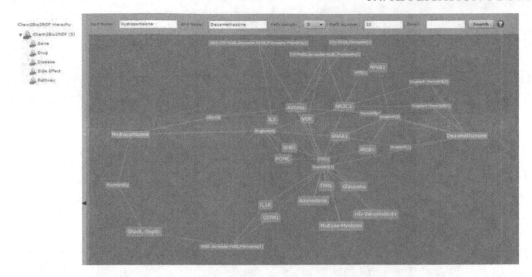

Figure 5.5: Search interface for association searcher.

1. The first part includes ranked paths and their calculated scores. The path score is the accumulation of association scores of all contained bio-term pairs. The association score for an individual pair is calculated based on the topic distribution of two bio-terms in the MEDLINE/PubMed abstracts, given by the Bio-LDA model. This ranking method will discover the paths with closer relations in the literature.

2. The second part contains the json.txt file, which is a special format to represent paths. It can be stored as a plain text file and placed in the same folder with the pathfinder.swf file, to visualize the flash result locally. The pathfinder.swf file can be downloaded from http://ch eminfov.informatics.indiana.edu/huiwang/pathfinder/file/pathfinder.swf.

5.4.3 ASSOCIATION EXPLORER

The association explorer is based on the methods mentioned in Section 5.4.3 (http://chemin fov.informatics.indiana.edu:8080/yuysun/radiuschembiospace.html). The interface is shown in Figure 5.6, and is similar to that of the association searcher, both based on Adobe Flash. The association explorer takes a start node, an end node type, the path length, and the number of path returns as required inputs. It first searches Chem2Bio2RDF for all qualified paths, and then ranks those paths using the KL divergence score from the Bio-LDA model. The top-ranked paths are displayed to the user.

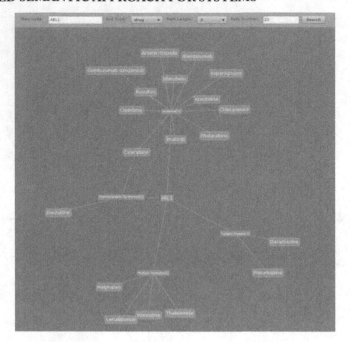

Figure 5.6: Search interface for association explorer.

5.5   USE CASES

The methods and tools discussed in previous sections are evaluated using case studies.

5.5.1   IDENTIFYING POTENTIAL DRUGS FOR A TARGET

As discussed in Section 5.3.1, bio-term associations are generated by combining the linked data resources (i.e., Chem2Bio2RDF) and literature resources using the Bio-LDA model. To illustrate, drugs that target the abelson murine leukemia viral oncogene homolog 1 (ABL1), which has been implicated in processes of cell differentiation, cell division, cell adhesion, and stress response, are investigated. ABL1 is also known as a factor in chronic myeloid leukemia. As shown in Figure 5.7, five drugs, cisplatin, adenesine triphosphate, imatinib, dasatinib, and gefitinib, all target ABL1, according to Drugbank (i.e., accounting for the solid lines in the diagram). Predicted drugs that may target ABL1 are generated via the gene-disease-drug association that must satisfy two conditions.

1. There exists a gene-disease-drug path in Chem2Bio2RDF.

2. The calculated association score for the gene-drug pair should be less than a certain threshold.

The association scores are computed using the Bio-LDA model, with 50 topics on 336,899 PubMed article abstracts published in 2009. The association scores based on the Bio-LDA model are given by Equation (4.8), which is also known as the symmetric KL divergence. A KL divergence score not larger than 5 is used as the threshold. Usually, there exist multiple gene-disease-drug paths in Chem2Bio2RDF for a given gene-drug pair. The accumulated score of each pair in the path is used to rank the possible paths, and only the one with the most significant scores will be shown in the network. The diseases leukemia, myeloma, and neoplasm are the most significant diseases that associate ABL1 with drugs. Figure 5.7 shows the generated network using the Bio-LDA model. 15 drugs are suggested by the Bio-LDA model. Similar to the directly linked five drugs, which are used in the treatment of various cancers, those predicted drugs are all chemotherapy related. The diseases leukemia and multiple myeloma are also highly associated with the ABL1 based on the analyses.

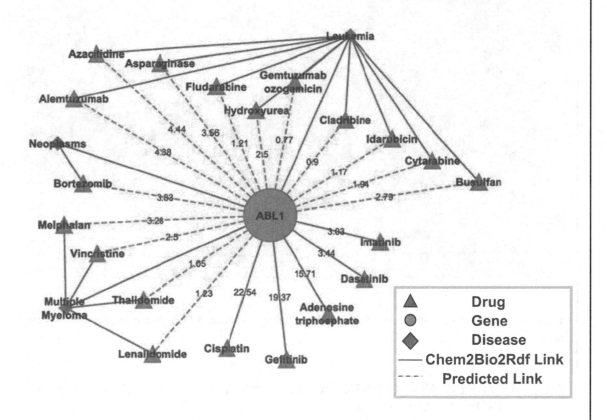

Figure 5.7: The association network for tyrosine-protein kinase ABL1 based on association prediction.

The association exploration method is also applied to ABL1 to identify the possible drugs, as shown in Figure 5.8. Although principles behind association prediction and association exploration are different, results for those two methods are very similar. Leukemia, multiple myeloma, etc., are identified as the most important connection nodes in both methods. The drugs showing high associations with ABL1 are also highly overlapped. Thus, users can choose either method to answer questions like *"what are the most relevant (drugs, genes, diseases, side effects, and pathways) for a known (drug, gene, disease, side effect and pathway)?"*

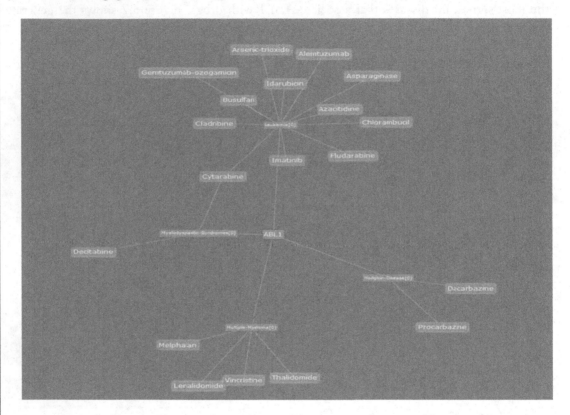

Figure 5.8: The association network for tyrosine-protein kinase ABL1, based on association exploration.

5.5.2   INVESTIGATING DRUG POLYPHARMACOLOGY USING ASSOCIATION SEARCH

In drug discovery, a major question is how to find drug candidates for a targeted disease. Since approximately 35% of known drugs have more than one target, the efficacy of many drugs is increasingly thought to come from their effect on multiple targets, which is known as *polyphar-*

macology. Based on this assumption, drug candidates can be identified from compounds that have the same multiple targets as a marketed drug. Thus, the question of how to find a drug candidate for a therapy can be formulated as a query in this system: to find all drug-like compounds that share at least two targets with a drug that is used for the therapy. For example, if a user wants to find some drug candidates for inflammatory and autoimmune conditions, such as rheumatoid arthritis, he can start with the typical drug, dexamethasone, and then search for compounds that have similar targets, with an activity score greater than 50 (activity score 0-100). The graphical representation of an example of the query process is shown in Figure 5.9.

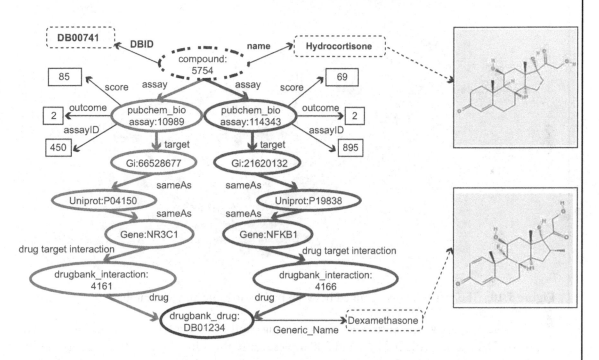

Figure 5.9: Graphic representation of the SPQRQL-query for finding a compound similar to dexamethasone.

To further understand the relation between the given drug, dexamethasone, and the found compound, hydrocortisone, the semantic association search proposed in Section 5.3.2 within Chem2Bio2RDF is explored. 47 near-shortest paths are found from hydrocortisone to dexamethasone, including five types, drug-gene-drug, drug-disease-drug, drug-gene-gene-drug, drug-gene-disease-drug, and drug-disease-gene-drug. These paths are then ranked based on the association scores calculated using Equation (4.8). The top ten paths and the association scores for each pair (based on 50 topics) are shown in Figure 5.10. Three similar gene targets, NR3C1, ANXA1, and NOS2, are shared by both dexamethasone and hydrocortisone. Among those paths, five paths are associated with NR3C1, glucocorticoid receptor, which indicates its significance in understanding the pharmacokinetics of drugs.

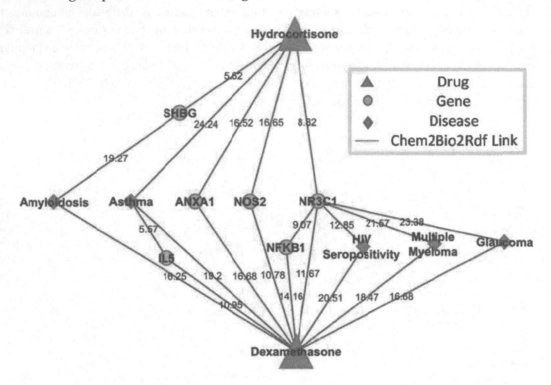

Figure 5.10: The top 10 paths obtained between hydrocortisone and dexamethasone.

Table 5.1 shows the entropy of the two drugs and three gene targets, calculated based on the Bio-LDA model with 50, 100, and 200 topics, using the 336,899 MEDLINE/PubMed article abstracts published in 2009, which contain 13,338 identical bio-terms. Here n represents the number of abstracts that contains given bio-terms in the literature set. Dexamethasone is a more effective drug than hydrocortisone, since it is involved in 742 more abstracts and has higher entropies. This makes sense from a biological point of view, as dexamethasone is 40 times as potent as hydrocortisone. Table 5.2 shows the symmetric KL divergence for pairs of bio-terms in this use case, and n shows the number of co-occurrence of the given bio-term pair. Hydrocortisone and dexamethasone co-occurred in 17 abstracts, and have lower KL divergence.

Hydrocortisone and dexamethasone both target genes NR3C1, ANXA1, and NOS2. For different numbers of topics (T=200, T=100, and T=50), Table 5.1 shows that the ascending order of the values of average entropies for the three genes is: ANXA1<NR3C1<NOS2, suggesting that NOS2 tends to be involved with more topics, while ANXA1 tends to be associated with fewer topics. Thus, the path between the two drugs with ANXA1 is more focused and specific,

Table 5.1: Bio-term entropies for nodes shown in the top three paths

Bio-Terms Name	Bio-Terms Identifier	Type	n	T = 200	T = 100	T = 50	Average
hydrocortisone	DB00741	Drug	139	2.558	1.880	2.454	2.297
dexamethasone	DB01234	Drug	881	4.292	3.754	3.484	3.843
ANXA1	P04083	Gene	23	2.266	1.631	1.365	1.754
NR3C1	P04150	Gene	16	2.123	2.840	2.486	2.483
NOS2	P35228	Gene	40	2.824	2.833	2.598	2.752

which intuitively conveys more contents. This makes sense as hydrocortisone and dexamethasone are involved in the de novo synthesis of the ANXA1 gene. Thus, the three paths involved with the three genes can be ranked according to their semantic specificity as: path with ANXA1> path with NOS2> path with NR3C1.

Moreover, the smaller the KL divergence of the path, the more semantically relevant the nodes and edges along the path. Table 5.2 shows that the entities and relationships along the path through NR3C1 are the most relevant to each of the three paths. Combining the entropy and KL divergence, it shows that the path with ANXA1 is more favorable in specific research, and the path with NR3C1 is more favorable in general research.

Table 5.2: Symmetric KL divergence for the top three paths

Bio-Term Semantic Associations	T = 200	T = 100	T = 50	Average
hydrocortisone~NR3C1~ dexamethasone	29.96	21.55	20.49	24.00
hydrocortisone ~NOS2~dexamethasone	35.40	31.00	27.42	31.28
hydrocortisone~ANXA1~ dexamethasone	43.39	40.31	33.20	38.97

The association search method has also been applied to several studies to help understand the mechanism behind drugs [He et al., 2011]. For example, this association search method is applied to find gene associations between thiazolinediones (e.g., rosiglitazone, troglitazone, and pioglitazone) and cardiac side effects. The comparison of association graphs between the thiazolinediones drugs and myocardial infarction identifies that ADIPOQ and APOE are the critical genes that are associated with cardio function recently reported in the literature [Bennet et al., 2007]. It also explains the success of pioglitazone in avoiding myocardial infarction, which is confirmed in recent clinical literature [Nissen and Wolski, 2007].

The study of associations among non-steroidal anti-inflammatory drugs (NSAIDs), inflammation, and Parkinson disease with the association search method indicates a strong connection between Ibuprofen and Parkinson's disease through hemorrhage and other genes, which is evidenced by recent studies [Dani et al., 2005, Ertel et al., 1992, Frazier et al., 2004, Pradilla et al., 2005].

5.5.3   BUILDING A DISEASE-SPECIFIC DRUG-PROTEIN CONNECTIVITY MAP

The molecular connectivity map shows how the expression level of genes changes in response to different drug compound perturbations, which enables researchers to compare the molecular therapeutic/toxicological profiles of many candidate drugs or drug-target genes, therefore improving the chance of developing high-quality drugs and reducing drug development time. In this study, a novel method is proposed to compute the high-coverage disease-specific drug-gene connectivity maps, by integrating chemogenomics sources (i.e., Chem2Bio2RDF) with literature hits from the Bio-LDA model. The purpose of the connectivity maps is to find novel therapeutic uses of old drugs, also known as drug repositioning.

Using Alzheimer's disease (AD) as an example, the gene list is created by searching for AD-related genes from the linked data (Chem2Bio2RDF). 88 genes are identified. 382 drugs are selected based on the drug-gene interaction. The gene list can be expanded to 13,998, and the drug list can be expanded to 1,898, if gene-gene interaction is involved. Experiments show that the top ranks of the expanded sets of genes and drugs are close to the lists without extension. Thus, gene-gene interaction is ignored, unless users specify its inclusion, to speed up the calculation.

The connectivity scores are calculated using the Bio-LDA association scores. Figure 5.11 shows the AD-related drug-protein connectivity map. The x-dimension represents drugs and the y-dimension represents genes. Hierarchical clustering of drugs and genes is performed using their Euclidean distances. The color intensity for each cell is drawn in proportion to the connectivity score, as shown in the heatmap legend. In the Bio-LDA model, the connectivity scores indicate the distance between the gene-drug pair. The smaller the score, the more significant the relationship. The cells with purple color indicate the most significant interactions related to Alzheimer's disease. From the figure and zoom-in boxes, the genes and drugs highly related to Alzheimer's disease are studied. For example, the CYP family is known to be highly associated with AD. The drug, ketoconazole, may affect some AD drug metabolism, such as that of donepezil. Diclofenac is a non-steroidal anti-inflammatory drug (NSAID). Research shows the NSAIDs may prevent the development of AD if given daily in small doses for many years.

5.5.4   ASSOCIATION SEARCH FOR DISCOVERY COMPOUNDS

All previous case studies required existing data in Chem2Bio2RDF. In drug discovery, project compounds are usually virtual compounds. The data related to these virtual compounds are very limited. In order to apply knowledge discovery methods to drug discovery, three new types of connections are used to append the current Chem2Bio2RDF: similarity connection, in-silico predicted connection, and experiment connection. The similarity connections are given by the Tanimoto-coefficient-based structure similarity search. The in-silico predicted connections are

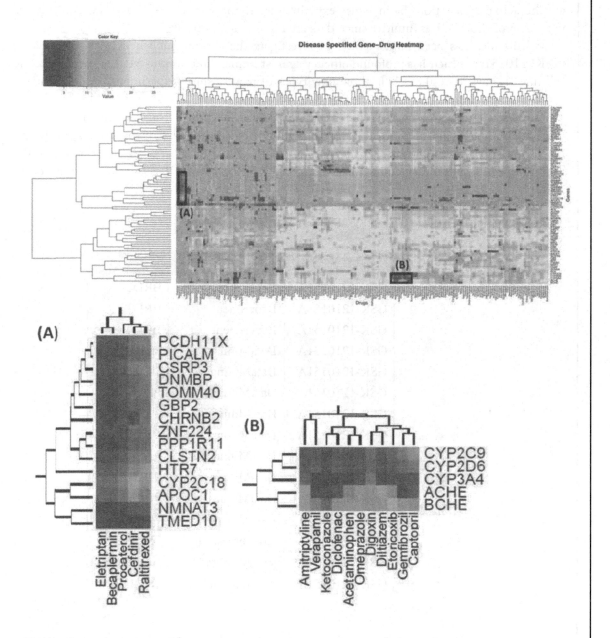

Figure 5.11: A connectivity map linking AD-related genes to significant drugs.

created based on the prediction model, using data mining methods. The experiment connections are the actual connections by in-house experiments. The integration of those new connections with Chem2Bio2RDF is intuitive since data are represented as triples.

This idea has been implemented to investigate the current published GVK compound, GSK1210151A, which has profound efficacy against human and murine MLL-fusion leukemic cell lines [Dawson et al., 2011]. Three new types of relationship, "experiment," "data mining," and "structure similar," are collected and added to Chem2Bio2RDF, as shown in Table 5.3. The experimental results are obtained from PROUS, which is recorded from patents filed by GSK. The data mining results are from a polypharmacology prediction model, using the support vector machine (SVM) method [Wale, 2011]. The structure similar drugs are calculated based on the Tanimoto coefficient measurement [Tanimoto, 1957], using 166 MDL public key [Durant et al., 2002].

Table 5.3: Added triples to Chem2Bio2RDF

Subject	Predicate	Object
GSK-1210151A	Experiment	BRD2
GSK-1210151A	Experiment	BRD3
GSK-1210151A	Experiment	BRD4
GSK-1210151A	DataMining	MAOB
GSK-1210151A	DataMining	CCR2
GSK-1210151A	DataMining	PDE5A
GSK-1210151A	DataMining	CYP2C9
GSK-1210151A	DataMining	NFKB1
GSK-1210151A	DataMining	PDE4C
GSK-1210151A	DataMining	PDE4A
GSK-1210151A	DataMining	PDE3A
GSK-1210151A	DataMining	PDE9A
GSK-1210151A	structureSimiliar	Micafungin
GSK-1210151A	structureSimiliar	Delavirdine
GSK-1210151A	structureSimiliar	Tadalafil

With the addition of new triples to Chem2Bio2RDF, it is possible to apply the association search between GSK1210151A and other bio-terms. Figure 5.12 shows the connections between GSK-1210151A and leukemia using the association search tool. Consistent with the published finding, bromodomain-containing protein 4 (BRD4) is one target that can affect leukemia through protein-protein interactions. Nuclear factor NF-kappa-B p105 subunit (NFKB1) is also identified in the graph that can affect leukemia cell lines through both direct target and indirect protein-protein interactions. This finding is consistent with several other independent studies, which found that NFKB1 is an important target for treating leukemia [Chang et al., 2006, Reuter et al., 2009].

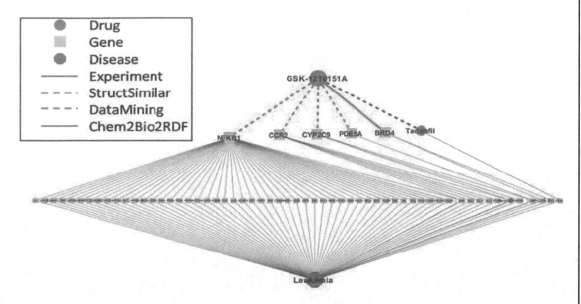

Figure 5.12: Paths between GSK-1210151A and leukemia.

5.6   CONCLUSION

In this chapter, the semantically connected data source is integrated with the data mining results,using the Bio-LDA model. Pathfinding algorithms are applied to the Chem2Bio2RDF network first, to identify all qualified paths. The obtained paths are then ranked by the association measurements, using the Bio-LDA model on the PubMed literature. Case studies are shown for the proposed, association prediction, association search, association explore, and connectivity map generation methods. These methods can contribute to crucial issues in the biomedical domain, including chemogenomics, polypharmacology, and drug repositioning. Expert and literature investigations are adopted to assess the result and value of the proposed algorithms, which indicates that the algorithms can help discover invisible knowledge and identify potential research issues by obtaining and integrating existing knowledge.

CHAPTER 6

Semantic Link Association Prediction

6.1   INTRODUCTION

Recent advances in the Semantic Web [Shadbolt et al., 2006] have enabled the creation of large heterogeneous networks of experimental data in life sciences, for example, Chem2Bio2RDF, LODD, Bio2RDF, OpenPHACTS, Linked Life Data, and Linked Open Data, where nodes can include physical and abstract entities (e.g., compounds, protein targets, substructures, side effects, diseases, pathways, tissues, and gene ontology terms), and edges (or links) represent various relations between objects (e.g., drug-drug interactions, drug-target interactions, and protein-protein interactions). The ability to easily integrate heterogeneous datasets in a meaningful way makes semantic technologies attractive, although it is only recently that supporting technologies have been adequately mature to make them useful in life sciences, in particular, the advent of fast triple stores for data storage, the SPARQL query language for searching, and the OWL ontology language for the description of ontologies. In contrast to hyperlinked data, semantic linked data encodes explicit meanings of nodes and links, making it possible to traverse from one node to another via particular kinds of relationships. Prediction of links not in the dataset, based on existing links, is widely used in social networking, in which it is assumed that two nodes are similar if they share common neighbors (e.g., a certain number of neighbors, and similar shortest paths). For example, in a co-authorship network, two authors are similar in terms of research interests if they co-author many papers together; hence, their potential collaboration could be predicted. It should be noted that social networks generally only deal with positive relationships; drug discovery data is different, in that negative relationships such as inactivity are important.

One critical topic of Systems Chemical Biology is to investigate drug-target interaction. Understanding the interaction of drugs with multiple targets can identify potential side effects and toxicities [Scheiber et al., 2009, Xie et al., 2007, 2009], as well as possible new applications of existing drugs [Ashburn and Thor, 2004, Dudley et al., 2011, Keiser et al., 2009, Kinnings et al., 2009, O'Connor and Roth, 2005]. Many efforts have been made to integrate drug-target interactions on a large scale [Garcia-Serna et al., 2010, Kuhn et al., 2010a, Oprea et al., 2011, Taboureau et al., 2011]. A variety of computational approaches have been previously explored for predicting drug-target interactions, including molecular docking [Li et al., 2011, Xie et al., 2009, Yang et al., 2011], ligand-based predictive models [Keiser et al., 2007, Nidhi et al., 2006], phenotype similarity, side effect similarity [Campillos et al., 2008] or gene expression profile similarity [Lamb et al., 2006], and chemical ontology similarity [Ferreira and Couto, 2010]. Some similarity measurements have been combined to elucidate drug targets [Perlman et al., 2011]. Network analysis based on the topology of known drug-target networks has also been utilized for drug-target prediction, but is currently limited to small datasets [Bleakley and Yamanishi, 2009, Zhao and Li, 2010]. The basic idea of the prediction methods is illustrated in Figure 6.1. Most of current studies are limited to individual factors, such as substructure; leveraging various information would be of great help for drug target prediction.

In this chapter, Semantic Web technologies were applied to integrate and annotate data related drug target interaction, constructing a heterogeneous network with over 290,000 nodes and 720,000 edges. A statistical model, called the Semantic Link Association Prediction (SLAP) model, was developed, to assess the association of drug target pairs and to predict missing links. An association score is calculated based on the topology and semantics of the neighborhood. Results demonstrate that SLAP can correctly identify known drug target pairs from random pairs with high accuracy, and can also identify indirect drug target relations (e.g., change of gene expression). The association scores of a drug against a set of targets constitute a biological signature that allows assessing the similarity of drugs in the context of a whole system. The resulting drug similarity network clusters contain drugs from the same therapeutic indication in ways not observed using chemical structure similarity, and can also be used to identify potential new indications for existing drugs. The algorithm can be extended to assess other types of association, such as drug-side effect, drug-disease, and target-disease.

6.2   MATERIALS AND METHODS

6.2.1   NETWORK BUILDING

We extracted drug-target interactions and data contributing to either the similarity of compounds, the similarity of targets, or chemical target interaction from Chem2Bio2RDF, and added semantic annotations using the Chem2Bio2OWL ontology, to create a semantic drug-target network. For example, two compounds are similar if they share same side effects, same substructures, or same chemical ontology terms; two targets are similar if they share the same gene ontology terms, ligands, or they function in the same pathway. A link between a drug and a target via bind type is established if there exists a binding affinity smaller than 30 um. Each node in the network is an instance of one of the classes. Figure 6.2 shows the workflow.

An ontology is used to annotate public datasets and integrate them into a semantic linked network. Two nodes are linked by one or more paths, but only a small number of significant paths are kept for association estimation. The path significance and drug target associations are assessed by statistical models derived from random samples.

Many of the relevant data sources are scattered around the Web and published in diverse formats. Some of them describe the same entities from different perspectives. The semantic relationship of these datasets to each other is often unclear. Semantic Web techniques offer an efficient way to annotate data and integrate them into a huge network. The network is presented

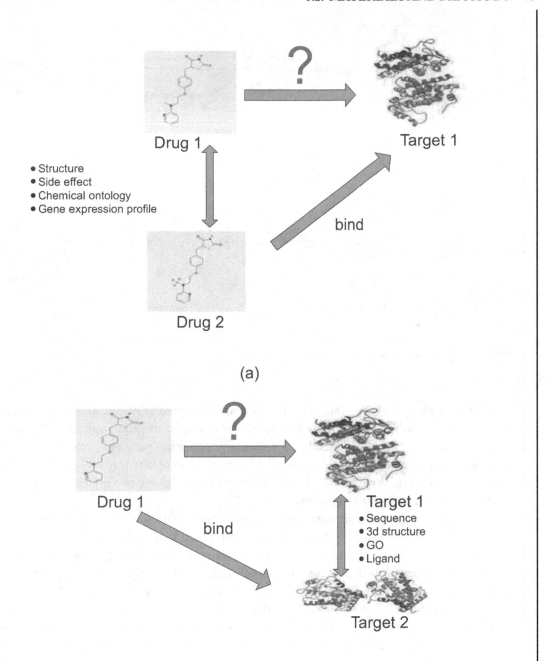

- Structure
- Side effect
- Chemical ontology
- Gene expression profile

(a)

Figure 6.1: Conventional drug target prediction approaches (a) from a ligand perspective and (b) from a target perspective.

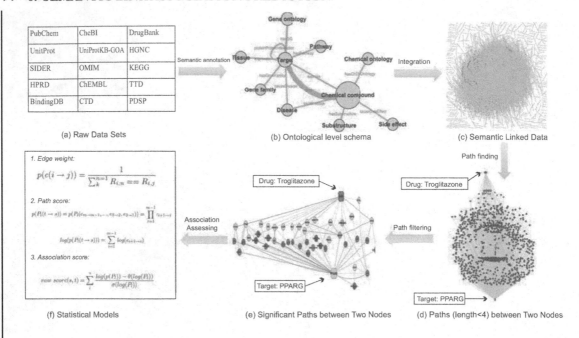

Figure 6.2: SLAP pipeline.

as a list of triples in which subjects and objects are presented as nodes, and predicates as edges. Each entity has a URI as a unique identifier. For example, the compound troglitazone is presented as `http://chem2bio2rdf.org/pubchem/resource/pubchem_compound/5591`. Each individual instance is mapped to a class. The major classes are listed in Table 3.1; they are linked by object properties. There may be different types of edges linking two nodes. For example, *Compound* could link to *target* by *binding*.

In data integration, all data have a unique agreed-upon identifier; other identifiers are mapped to the unique identifier by in-house scripts and manual inspection. The general integration procedure was explained in the previous chapters. Some changes for this particular case are pointed out here. We used expression data from CTD (by searching the interaction type ='expression') for express type. The binding type requires the activity to be less than 30um if it exists. GO terms were filtered out by using qualifier!='NOT' and evidence!='NAS' and evidence!='ND' and evidence!='NR' [Rhee et al., 2008]. We noticed that most of the edge types (e.g., protein-protein interaction and drug-target binding) conform to a scale-free property, in which degree distribution follows a power law distribution; but some promiscuous nodes (with many neighbors) skewed the distribution. To name a few: GO:0005515 (molecular function) in the GO class; CHEBI:25700 (organic molecular entity) in the Chemical Ontology Class; C0027497(Nausea) in the Side Effect Class, Aromatic compounds in the Substructure class, and Brain in the Tissue classes. These highly promiscuous nodes were removed manually from the network. The whole dataset is available at `http://chem2bio2rdf/slap`.

6.2.2 DRUG TARGET PAIR PREPARATION

Drug target pairs from DrugBank were used to build the network. We only took the pairs in which drugs were small molecules (i.e., by mapping to PubChem) and targets were homo sapiens (i.e., by mapping to HGNC). A total of 5,607 pairs were extracted from the network as one benchmark dataset for model evaluation. The drug target pairs were grouped into six classes according to ChEMBL [Gaulton et al., 2012] target classification: enzyme (2393 pairs), membrane receptor (862 pairs), ion channel (392 pairs), transporter (209 pairs), transcription factor (208 pairs), and others (1543 pairs). Another benchmark dataset was created from MATADOR, which was not used for network building. We took drug target pairs with direct interaction types and confidence score > 800 from MATADOR. 1,176 direct pairs in MATADOR were used, in which 1,065 pairs had at least one path with length $l \leq 3.3665$ indirect pairs in MATADOR were also extracted for evaluating indirect drug target interaction. Indirect interactions are caused by many different mechanisms, such as binding with drug metabolites or changing gene expressions.

6.2.3 PATH FINDING

A heap-based Dijkstra algorithm was employed to quickly find the paths between two nodes [He et al., 2011, Wang et al., 2011]. It can achieve a complexity of *O(nlogn)*. Each path is represented as: *node 1 - edge 1 - node 2 - edge 2 - · · · · - node n*. The length of a path is the number of edges between two nodes. We only took the paths of length $l \leq 3$. Only significant paths (assessed by statistical models) are visualized in Cytoscape [Shannon et al., 2003].

6.2.4 STATISTICAL MODEL

We randomly sampled 100,000 drug target pairs from DrugBank, covering 1,355 approved small molecule drugs and 1,683 human targets, and found 54,414 pairs with at least one shortest path with length $l \leq 3$. The sampling yielded 2,344,026 paths, which were categorized into 34 path patterns. The scores of each pattern were fitted to a normal distribution (Figure 6.3) and the expected mean and standard deviation were estimated, followed by calculation of the z score of every path. Only the paths with z score greater than 0 were considered as valid paths contributing to the association. The z scores of all the valid paths from s to t were summed to get an association score, which was later used to measure the strength of the association:

$$ raw\ score\,(s, t) = \sum_{l}^{n} \frac{\log\left(p\left(P_l\right)\right) - \theta\left(\log\left(P_l\right)\right)}{\sigma\left(\log\left(P_l\right)\right)}, $$

where $\log(p(\log(P_l))) > \theta(\log(P_l))$; n is the number of shortest paths between the nodes s and t; $\theta(\log(P_l))$ and $\sigma(\log(P_l))$ are the expected mean and expected standard deviation of the pattern to which P_l belongs.

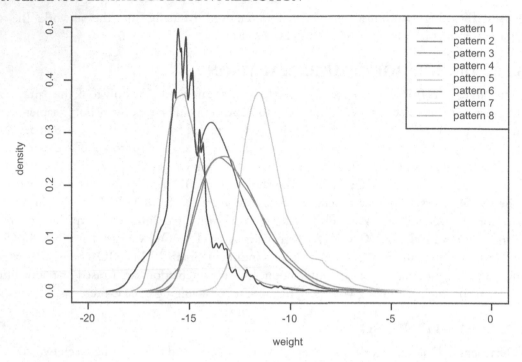

Figure 6.3: Raw score distribution of eight path patterns.

Some patterns may not be helpful or might even be noisy for assessing drug target association. We built a test set consisting of drug target pairs from DrugBank, and the same number of random drug target pairs sampled from the set of drugs and targets composing of the real drug target pairs. For each pair, raw scores of all the paths within a path pattern were calculated and summed up as a score for that path pattern. The scores were then used to rank the pairs in the test set. The evaluation of each pattern was performed using the area under ROC. We also applied the same procedure to the direct pairs from MATADOR. The patterns with low ROC ($AU\ ROC$ <0.51) were considered uninformative. The patterns that occurred as uninformative in both test sets, taken from DrugBank and MATADOR, were removed.

The logarithmic association scores of random pairs conform to a normal distribution (Figure 6.4); p-value is estimated to show the probability of observing a given score by random chance alone. A lower p-value indicates a stronger relation between two objects.

6.2.5 MODEL EVALUATION

A test set was composed of a set of drug target pairs from DrugBank, and the same number of random pairs as decoys. Three other test sets were created, by increasing the number of random

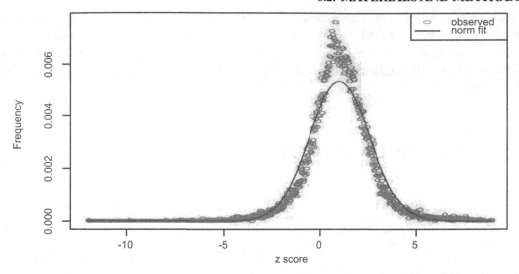

Figure 6.4: Fit association scores of random pairs to a normal distribution (Logarithm is applied to the scores. R2 is 0.96).

pairs such that the sizes of the random pair sets are 4, 8, and 12 times the size of the true drug target pairs. For each pair, the paths including the direct link (if found to exist) were removed, and the z scores of all valid paths were summed up as the association score. The scores were ranked to generate ROC curves [Fawcett, 2006], which are widely used to measure drug target prediction methods [Jacob and Vert, 2008, Perlman et al., 2011, Yamanishi et al., 2010, Zhao and Li, 2010]. We also considered the Precision and Recall (PR) curve, which shows the ratio of true positives among all the predicted positives under a given recall rate [Davis and Goadrich, 2006]. The PR curve is more informative and biologically meaningful when the dataset is unbalanced. The same procedure was also applied to another dataset collected from MATADOR. Other than using SLAP scores, we considered the number of shortest paths (maximum length 3), the number of valid paths (i.e., significant paths defined in the model), the sum of the raw scores of all paths, the maximum raw score among all paths, and the average raw score of all paths. In addition, we took the pairs validated in experiments in a recent published paper [Keiser et al., 2009] as novel pairs, after manually mapping their drugs and targets to PubChem CIDs and gene symbols; then we ran SLAP to get p-values of all the valid pairs.

6.2.6 ASSESS DRUG SIMILARITY

We identified drug-disease pairs from Yildirim et al. [2007], and then mapped the drugs to Pub-Chem CIDs. Many drugs have multiple indications, so in order to visualize drugs by therapeutic indications, only drugs with one indication were kept. We also only kept the top ten diseases, ordered by the number of related drugs. The association scores of all mapped drugs against a set of human targets construct biological signatures, which were later used for measuring drug similarity using the Pearson correlation coefficient. The pairs with coefficient $r > 0.9$ constitute the network. Drug structural similarity was measured by the Tanimoto coefficient, using the MACCS fingerprint.

6.3   RESULTS

6.3.1   SEMANTIC LINKED DATA

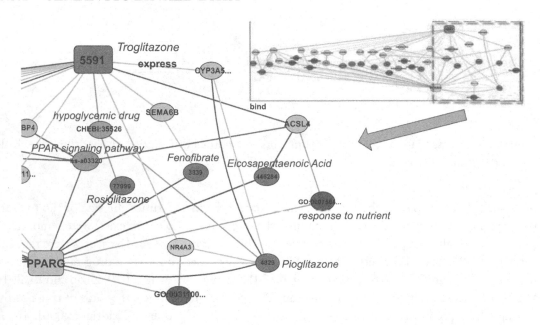

Figure 6.5: Paths between troglitazone (labeled as PubChem ID: 5591) and PPARG with length $l \leq 3$. Note: The nodes and edges are colored by their classes and edge types, respectively. Some nodes are annotated for clarity.

The SLAP pipeline is shown in Figure 6.2. A heterogeneous network consisting of 295,897 nodes and 727,997 edges was constructed from 17 public data sources pertaining to drug target interaction. Every node and edge was semantically annotated using Chem2Bio2OWL. The nodes were grouped into ten classes, which are linked by 12 types (Figure 6.2b). A single node is an instance of a corresponding class; for example, a node for the drug troglitazone (labeled as 5591 in Figure 6.5) is an instance of class Chemical Compound. We group paths of nodes and edges that share the same semantics (but with different instances) as path patterns, each path is an instance of a path pattern. Table 6.1 shows six path pattern examples between Drugs and Targets. In Figure 6.5, the path from node 5591 (troglitazone) to node PPARG (glitazone receptor) via

ACSL4 (long-chain-fatty-acid CoA ligase 4) and 446284 (eicosapentaenoic acid) is an instance of path pattern 1 in Table 6.1. We can interpret this path as indicating troglitazone could bind to ACSL4, which shares compound eicosapentaenoic acid with target PPARG. With the assumption that two nodes are associated if they link to at least one other node, or their linked nodes are linked, their relations can be assessed by the analysis of the links (or paths) between the two nodes [Liben-Nowell and Kleinberg, 2007]. The strength of their relation in the network can be measured by the distance, the number of shortest paths, and other topological properties between the two nodes. In our example of the relationship between troglitazone and PPARG, several paths provide "evidence" of a relationship: troglitazone and rosiglitazone both are hypoglycemic drugs and the latter is the ligand of PPARG; and troglitazone binds to ACSL4, which shares pathway (PPAR signaling pathway), ligand (eicosapentaenoic acid) and GO term (response to nutrient) with PPARG. A total of 1,684 paths (length $l \leq 3$) belonging to ten path patterns contribute to this relation.

6.3.2 PATTERN SCORE DISTRIBUTION

Each path between two nodes may contribute to the relation between them, but the degree of contribution varies, depending on path distance and the weight of edges involved in the path. For example, a gene ontology molecular function term (GO:0005515) shared by proteins is not as informative as a binding term (GO:0005488) in assessing the similarity of two proteins. Thus, the weight of the edge linking one protein node to the molecular function node is lower than that linking to the binding node. Based on this observation, we developed a statistical model to measure the weight of edges, as well as the significance of paths. The model takes into account the distance and the weight of each edge, and renders a raw score indicating the strength of each path. We found that the raw scores within the same path pattern are normally distributed, while the mean and standard deviation of patterns are different (Figure 6.3). Z scores converted from raw scores based on the pattern-score distribution are used to measure the contribution to the association: the higher the z score, the greater the contribution of the path. The sum of the z scores of all paths is defined as the association score, indicating the association strength of the drug target pair. The logarithm of association scores of random drug target pairs fit a normal distribution (Figure 6.4), that enables calculation of the significance of a given association score. For our troglitazone and PPARG example, the p-value is 9.06E-6, indicating a strong association.

6.3.3 PATTERN IMPORTANCE

A low p-value in a drug-target pair indicates a strong probability of association between the drug and the target, but it does not necessarily mean the drug and the target would interact biologically. Some patterns may be uninformative. We therefore considered each pattern as a feature, and assessed each feature by itself for its ability to identify drug-target pairs from random pairs across the set. Table 6.1 lists three informative patterns and three uninformative patterns, along with ROC scores. The first two patterns illustrate that the drug likely interacts with a protein that shares commonalities in terms of GO or ligand-binding profile with an existing target that the drug is already known to interact with. The third pattern indicates that the drug likely interacts with a protein with which another structurally similar drug could interact. As a result of this analysis, 12 "uninformative" patterns were removed. The sum of the z scores of a given pair is the sum of the z scores of the paths belonging to the informative patterns.

Table 6.1: Path pattern examples

Path Patterns	AUROC
Chemical/Drug–*bind*–Target–*bind*–Chemical/Drug–*bind*–Target	0.850
Chemical/Drug–*bind*–Target–*hasGo*–GO–*hasGO*–Target	0.824
Chemical/Drug–*hasSubstructure*–SubStructure–*hasSubstructure*–Chemical/Drug–*bind*–Target	0.620
Chemical/Drug–*express*–Target–*hasPathway*–Pathway–*hasPathway*–Target	0.495
Chemical/Drug–*express*–Target–*hasTissue*–Tissue–*hasTissue*–Target	0.501
Chemical/Drug–*express*–Target–*PPI*–Target	0.501

Edge types are presented in italics. AUROC shows the performance of predicting drug target interaction with the pattern alone. The first three patterns are more informative than the last three in their capability to contribute to the associations.

6.3.4 ASSOCIATION SCORES OF DRUG TARGET PAIRS

We randomly selected 1,000 known drug target pairs from DrugBank and compared their association scores with 1,000 random pairs of drugs and targets sampled from DrugBank. For each drug target pair, their direct link was removed in the score calculation so that their association was only determined by their neighborhood properties. We thus aimed to test the ability of SLAP to correctly identify "missing links" in the data, with the assumption that this might be used, for instance, to profile a group of compounds against an identified set of targets. As Figure 6.6 shows, random pairs have a broad range of scores, but most of them are close to zero. Overall, real drug-target pairs have much higher scores than random pairs ($p < 2.2E - 16$, using *paired-t* test). We also took all drug target pairs from DrugBank (in total 5,607 pairs in which 4,508 pairs have at least one path with length $l \leq 3$). We sampled the same number of random drug target pairs as decoys, to check the capability of identifying real drug target pairs by SLAP. We compared SLAP with other link prediction methods adopted in social network analysis [Liben-Nowell and Kleinberg, 2007]. The AUROC of SLAP is 0.92, outperforming other methods (i.e., the number of shortest paths, and the number of valid paths) (Figure 6.7). As the ratio between true drug target pairs versus random pairs decreases (e.g., ratio=1/12), the ROC scores do not vary much ($AUROC \approx 0.92$) and SLAP still performs much better than others, although the precision goes down considerably (Figure 6.8). Even when random pairs are 12 times as much as

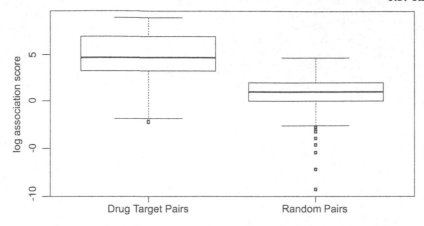

Figure 6.6: Logarithmic association score distribution of drug target pairs.

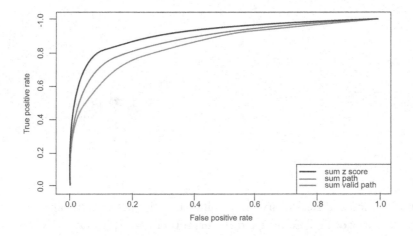

Figure 6.7: ROC curves among different prediction methods (Valid paths mean their z score > 0).

positive pairs, the precision still can reach 0.6 while recall is 0.7. In addition, we noticed using the sum (or max or mean) of raw score of the shortest path (without converting into z scores) performs as a random choice, indicating the importance of introducing random samples. Since several drug target prediction approaches reported that the performances may vary among different target classes [Yamanishi et al., 2010], we grouped the drug target pairs into 5 classes (e.g., Enzyme, Membrane Receptor, Ion Channel, Transporter, and Transcription Factor), and found that the score does not have any preference to a particular target class, indicating SLAP is capable of treating different classes of protein targets (Figure 6.9).

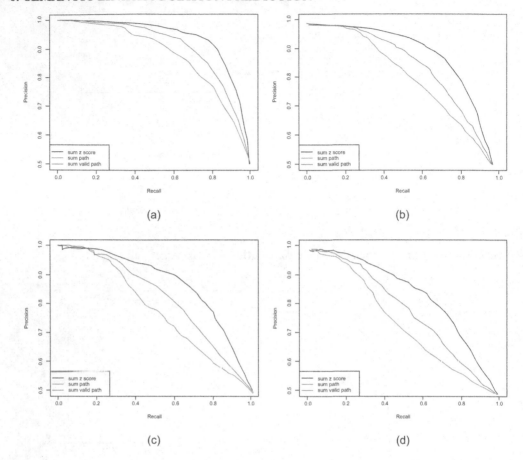

Figure 6.8: Precision and Recall curve under different ratios between the number of true drug target pairs and the number of random drug target pairs ((a) ratio=1:1, (b) ratio=1:4, (c) ratio=1:8, and (d) ratio=1:12).

As far as we are aware, SLAP is the only network predictive model on large scale graphs that has been applied to the drug discovery data. However, other drug-target prediction methods have been the subject of recent publications [Campillos et al., 2008, Keiser et al., 2009, Vidal and Mestres, 2010], and we thus sought to consider how the effectiveness of SLAP compares with these methods. We ran SLAP against 23 drug target pairs (including 15 aminergic G-protein-coupled receptors and 8 cross-boundary targets) predicted and confirmed in using the SEA method [Keiser et al., 2009], a novel drug prediction method based on similarity analysis. Nine pairs of aminergic GPCRs were identified by SLAP ($p < 0.05$); one pair was not decided ($p > 0.05$); the rest of GPCRs have no mappings in the network (i.e., the drug was not found in

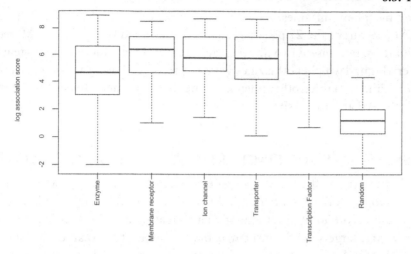

Figure 6.9: Logarithmic association scores of pairs among five gene families and random pairs.

the network), while only one of eight cross-boundary targets was identified by SLAP, indicating that SLAP is not capable of finding surprising pairs (e.g., cross-boundary targets). For example, vadilex, an ion-channel drug, was predicted in SEA as a ligand of a transporter, a totally different target, but was not identified by SLAP. Nevertheless, SLAP performs fairly well among GPCRs in this case.

In addition, we examined drug target pairs from MATADOR, which served as an external dataset for validation. 1,065 direct pairs were collected, of which 444 pairings are not represented in our network. 560 out of 621 known pairs and 170 out of 444 unknown drug target pairs were identified by SLAP ($p < 0.05$).

6.3.5   COMPARISON WITH CONNECTIVITY MAPS

By calculating association scores across multiple targets, SLAP can be used to build a polypharmacology profile of a drug even when a full data matrix is not available from drug-target experiments. We took all the 164 small molecules from the Connectivity Map (CMap), an online dataset mapping relationships of disease profiles to known drugs [Lamb et al., 2006], and 113 molecules that were mapped to our network were used to build a library. The association scores of these compounds against 1,683 targets were calculated, yielding a $113 \times 1,683$ score matrix. The targets for which the maximum score was smaller than 113 ($p < 0.01$) were eliminated, so that each remaining protein is a target of at least one drug. After this filtering, a matrix composed of 113 compounds and 679 targets was built. We used the signature of a given drug to compare it with all the compounds in the library, to find the most similar drugs, according to Pearson correlation coefficient. Following the CMap approach, eight queries, including two HDAC inhibitors, one estrogen and five phenothiazines, were created. We set 0.75 as a threshold. 21 pairs were identified by SLAP; 19 out of the 21 pairs were actually identified by CMap. SLAP recovered all the HDAC inhibitors, but missed two hits (i.e., genistein and tamoxifen) for estrogen; however, both hits rank very high. Two phenothiazines were not recovered using this similarity threshold, but they are quite similar to three other phenothiazines in the library. The results show that most hits identified by SLAP are true positives, indicating that profiles derived from SLAP resemble gene expression profiles being used for target identification.

6.3.6 ASSESSING DRUG SIMILARITY FROM BIOLOGICAL FUNCTION

We took 157 drugs from 10 disease areas to determine whether SLAP is able to distinguish drugs from different therapeutic areas. For each drug, we ran SLAP against 1,683 human targets and got an association score for each drug target pair, creating a $157 \times 1,683$ score matrix. We only kept the drugs and targets in which the maximum score is at least larger than 113 ($p < 0.01$) to make sure each drug has at least one valid target and each target has at least one valid drug. The matrix was then reduced to 147×339, followed by the correlation calculation of every drug pair. Only pairs with coefficient $r > 0.9$ were taken to build a network. A drug-similarity network was created accordingly (see Figure 6.10), which serves the following functions.

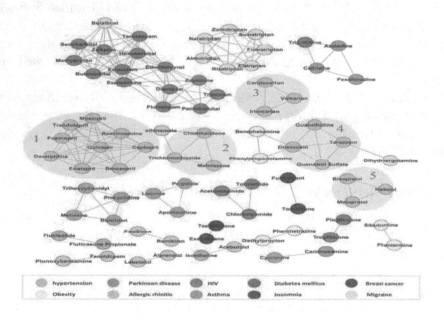

Figure 6.10: Drug similarity network. Note: Each node presents a drug, and two nodes are linked if their similarity (in terms of polypharmacology profile) $r > 0.9$. The drugs are colored by their therapeutic indication. Five hypertension-related clusters are shadowed.

1. Identifying mechanisms of action: Drugs with the same therapeutic indication tend to cluster together (Figure 6.10). For example, hypertension drugs subcluster into ACE inhibitors, thiazide-based diuretics, angiotensin II antagonists, alphaadrenoreceptor antagonists, and beta blockers (clusters 1–5 in Figure 6.10, respectively).

2. Calculating similarity of drugs by biological function: Mostly, chemically similar drugs have similar biological functions. However, small changes of structure may also result in big changes of function, or even totally different indications. For example, adding a methyl group to levodopa, a dopaminergic agent for Parkinson's disease, makes it methyldopa, an antiadrenergic (Tanimoto coefficient=0.89; Figure 6.11b) for antihypertension. They are distinguished by SLAP (similarity TC < 0.3). The antihypertensive effect of methyldopa is likely due to its metabolism to alpha-methylnorepinephrine (CID:3917). SLAP is still able to distinguish its metabolite from levodopa (similarity TC=0.23). Conversely, biologically similar drugs identified by SLAP are not necessarily structurally similar. For example, a number of drugs treating insomnia are quite different in term of structure (Figure 6.11a), but are clustered together by SLAP.

3. Drug repurposing: Some drugs with very different indications are clustered together. This may suggest some new indications of drugs or possible side effect considerations. For example, butalbital, a barbiturate used to treat migraines, is clustered with nine insomnia drugs, two of which (butabarbital and secobarbital) are barbiturates. Barbiturates act as central nervous system depressants, capable of producing all levels of CNS mood alteration, including insomnia. Triprolidine, an HIV drug, is a first-generation histamine H1 antagonist used in allergic rhinitis (and is clustered with other rhinitis drugs). Cycrimine is a central anticholinergic drug, designed to reduce the levels of acetylcholine in the treatment of Parkinson's disease, while its neighbor, carbinoxamine, used for allergic rhinitis, is likely capable of treating mild cases of Parkinson's disease as well `http://www.ebi.ac.uk/chebi/searchId.do?chebiId=3398`. It should be noted that since SLAP does not differentiate positive and negative interactions (i.e., activation or inhibition), the pairs may present opposite indications. Phenylpropanolamine (i.e., an alpha-1A adrenergic receptor agonist), clustered with doxazosin (i.e., an alpha-1A adrenergic receptor antagonist for treating hypertension) is known to cause severe hypertension [Pentel et al., 1985].

6.4   WEB SERVICES

SLAP was implemented as a Web service (`http://chem2bio2rdf.org/slap`)), with which we are able to perform: (1) one drug target pair association prediction and network exploration; (2) drug target prediction against hundreds of proteins; (3) prediction of target ligands; (4) large-scale prediction of drug target pairs; (5) identification of similar drugs based on biological func-

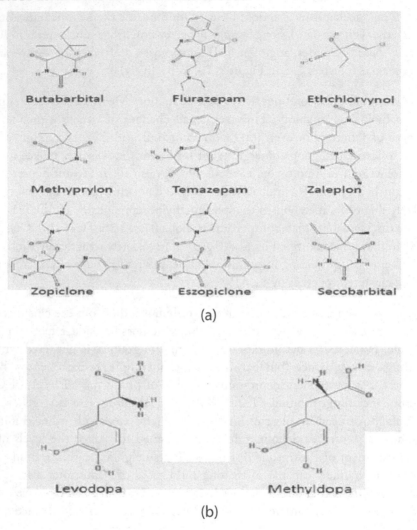

(a)

(b)

Figure 6.11: Comparison by structure similarity ((a) sample insomnia-related drugs and (b) levodopa vs. methyldopa).

tions; and (6) drug target prediction using other predictive models. Figure 6.12 shows the main interface of the website. It provides the following functionalities.

1. It supports chemical name, SMILES, and PubChem CID for compound/drug input, and supports gene symbol, protein name, and UniProt ID for target input. SMILES can be generated from an embedded molecular editor. For a new compound, the most similar

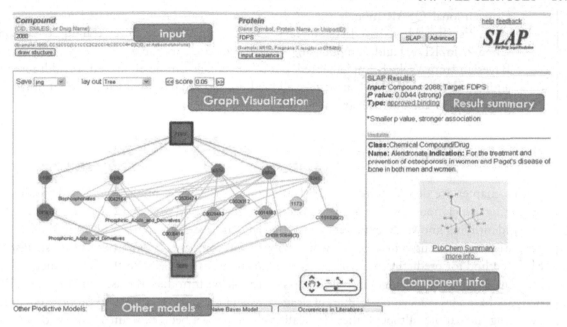

Figure 6.12: SLAP Web service.

compound (TC<0.85) is automatically retrieved and substituted in the prediction. For a new target, the most-similar target can be retrieved, based on sequence similarity.

2. The P value, used to measure association, is shown in the result summary window. A smaller P value indicates a stronger association. A pair with a P value <0.05 is considered significant. If such a pair exists in the system, its original RDF set published in Chem2Bio2RDF will be cited.

3. The network can be explored in an exploratory window. The nodes and edges are colored based on their semantic meanings. Paths can be filtered by their contribution to the association. Nodes with the same semantics and the same neighbors are merged into one virtual node, and the number of nodes being merged is labeled on the virtual node. Several options for network layout and export format are offered.

4. When you click a node or edge in the network, its basic information is retrieved from the Chem2Bio2RDF SPARQL endpoint and other resources, and is shown in the information window. The links to the original RDF sources are shown as well. For example, the "evidence" link of the edge refers to the sources curating this relation.

5. Conventional QSAR models (i.e., *Naive Bayesian*) can be implemented in the system via APIs. The AJAX-based system calls a Web service that retrieves training data from Chem2Bio2RDF, builds a model, and makes predictions on the fly. Currently, only Naive Bayesian from Weka is implemented, but other models can be employed. In addition, we added a Web service that can retrieve all the PubMed articles in which the input drug and target both occur in the abstract.

6.5   DISCUSSION

In this chapter, we demonstrate the SLAP method of association prediction, and the utility of predicting associations based on semantic networks. The method performs extremely well in correctly identifying known drug-target pairs in the data, has been shown to outperform similar link prediction methods used in social network analysis, and compares favorably with the established SEA method for predicting new drug-target interactions, as well as with the CMap method for associating drugs with changes in gene expression levels. We introduce the use of a drug-similarity network based on association profiles of drugs across targets, and use these to propose potential new drug indications, although these indications have not yet been experimentally validated.

The use of large semantically annotated datasets to identify potential relationships in the linked data is a new area. There are several limitations to our current version. First, adding more data pertaining to drugs and targets would help identify related pairs. The side effect, disease, and chemical ontology data are only linked to a limited number of drugs at present, and protein-protein interaction and protein pathway mapping data should greatly enhance its utility. In particular, the ability to embed compounds into the network for which there is no public information, using chemical structure similarity, or new targets into the network, using sequence similarity, would enable predictions to be made (albeit more indirectly) for newly synthesized or resolved compounds and targets. Second, as the complexity of pathfinding increases dramatically with increasing path length, only the shortest paths (i.e., length $l \leq 3$) were considered, thus potentially missing important path patterns that have a greater path length. Third, edge weights are defined with the assumption that the probability from one node to its neighbors with same semantic type (e.g., from one drug to its targets) is equal. An important limitation of our current algorithm is that it does not enable differentiation of relationships, other than categorical ones defined in the ontology. For instance, binding affinity could be used to weight the edge between drug and target; the edge with lower affinity is expected to have higher probability than that with higher affinity (or inactive interaction). Using such data brings up the issue of comparability between datasets: Some chemogenomics datasets (e.g., DrugBank) currently do not provide sufficient binding affinities, but the weighting schema can be modified straightforwardly in SLAP once the data are provided. In addition, binding types (i.e., agonist/antagonist, activator/inhibitor) can be incorporated to classify edges. Fourth, it should be pointed out that using large publicly integrated datasets means there is often a fuzziness between "no data" and "inactive data": for example, we cannot assume that because two items do not have a relationship in the dataset, that they are not related.

A key question in employing any drug-target prediction method is the extent to which it requires data completeness—in the extreme, a full experimental matrix—to work properly (i.e., if it needs to be trained with consistent known active/inactive information for all compounds against all targets). Our methods do not require such training; indeed their purpose is to suggest potential "missing links" in incomplete data. However, it should be pointed out that the level of data completeness in a set will affect the path lengths, z-scores, and association scores produced. We believe that, overall, SLAP should be considered a useful tool for predicting that a relationship exists between drugs and targets, and thus as a tool primarily for hypothesis generation, and for suggesting relationships to be probed experimentally. Its purpose is to predict a relationship, not necessarily indicating a strong physical interaction. We believe it is also useful, as demonstrated in our drug network, for profiling compounds by their target associations (and vice versa), and we plan to explore other types of networks that can be derived from SLAP.

Many drug target prediction methods only employ a single kind of information or relationship (e.g., substructure, side effect, etc.); these methods are limited due to incompleteness of the data, for instance drug target relations are far from complete [Mestres et al., 2008]. The employment of a variety of data can compensate for the lack of completeness of individual information categories. SLAP shows a way to leverage such information for drug target prediction. Several sample pairs, along with their key information, are listed in Table 6.2. For instance, the association between pyridoxal phosphate (CID: 1051) and cysteine conjugate betalyase 2 (CCBL2,) is very strong (p-value=1.9E−3); but if we remove gene ontology information, their association becomes very weak (p-value=0.02). The association between dexamethasone (CID: 5743) and annexin A1(ANXA1) would hardly be captured if substructure information were not considered.

Table 6.2: Sample drug target pairs with/without key information contributing to the association

CID	Drug Name	Gene Symbol	Target Name	Key Information	p value without key information	p value with key information
6741	methylprednisolone	NR3C1	glucocorticoid receptor	Chemical Ontology, Expression	0.03	8.0E-4
5743	dexamethasone	ANXA1	annexin A1	substructure	0.15	3.8E-3
8223	ergotamine	ADRA1A	adrenergic, alpha-1A-, receptor	ligand	0.09	4.0E-5
1051	pyridoxal phosphate	CCBL2	cysteine conjugate-beta lyase 2	Gene ontology	0.02	1.9E-3

The most compelling advantage of SLAP is its consideration of relations from a system level, rather than just from the known binding affinity data. Other than direct drug target interactions, SLAP is also capable of recognizing indirect interactions (e.g., change of gene expression level) from random pairs, although the association scores are often smaller than direct interactions (Figure 6.13). It thus allows us to evaluate drug similarity based on biological function. The network demonstrates that such similarity measurements are not only able to identify the drug action modes, but could also suggest new uses of drugs.

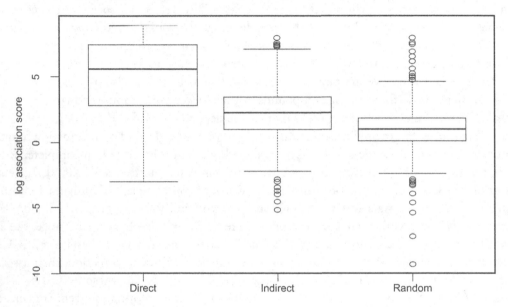

Figure 6.13: Logarithmic association scores of direct drug target pairs vs. indirect pairs. Indirect pairs were taken from MATADOR.

CHAPTER 7

Conclusions

In this book, we introduced the new concept "Semantic Systems Chemical Biology" to emphasize the importance of bridging together two research areas: the Semantic Web and Systems Chemical Biology. These two areas were articulated in Chapter 1, where the semantic techniques were surveyed, and the main challenges of bringing Semantic Web into Systems Chemical Biology were discussed. In the following chapters, we presented four major works that led to the building of a framework of Semantic Systems Chemical Biology.

Chapter 2 proposed a semantic framework using RDF, in which all public chemogenomics datasets were converted into RDF triples, filling the gap between LODD (with emphasis on drug data) and Bio2RDF (with emphasis on biological data). We further demonstrated the potential of linking multiples datasets via RDF by examining several key problems in Systems Chemical Biology.

We found that using RDF alone is not sufficient to model the data, limiting its application in addressing complicated questions in an intuitive manner; thus in Chapter 3, we introduced Chem2Bio2OWL, an OWL ontology whose aim is to integrate data at the concept/entity level, rather than at the database level. The benefits of concept-based integration were demonstrated by several use cases, including the facilitation of data retrieval and network-based mining.

Chapters 4 and 5 developed novel knowledge discovery approaches, including data mining, text mining, and Semantic Web technologies, to remediate the current information-explosion challenge in the pharmaceutical industry. Specifically, this work demonstrated the development of a data mining approach for compound pre-selection, a text mining approach for literature retrieval, and a Semantic Web approach for data integration. We applied a novel Bio-LDA model to PubMed, using an extension to Latent Dirichlet Allocation (LDA) and a set of semantically associated terms called bio-terms. The Bio-LDA model produces accurate latent topics and identifies quantitative associations between biomedical entities. Associations identified by Bio-LDA are combined with paths generated from Chem2Bio2RDF, to provide additional insightful predictions. This combined approach of Semantic Web and text mining approaches provides new opportunities for large-scale chemical and biological knowledge discovery.

In Chapter 6, we showed a novel statistical model based on semantics using linked data, and applied this model to assessing drug target association, one of the most challenging topics in the current drug discovery process. This work demonstrated the benefits of leveraging multiple heterogeneous sources, contributing to the drug target association via the application of Semantic Web technologies, which conventional models usually failed. Finally, the large-scale prediction of drug targets was applied, to investigate the mechanism of action of drugs and repurpose existing drugs.

In summary, this book covers techniques for building a semantic framework to study Systems Chemical Biology, ranging from data representation and integration to data mining. The case studies demonstrate the potential of bringing Semantic Web technologies and statistical methods into the study of Systems Chemical Biology. The large-scale application of these methods to drug discovery will greatly benefit from the massive development and adoption of domain ontologies, the building of more user-friendly SPARQL endpoints, more sophisticated visualization tools, and more efficient triple stores for data retrieval and traversal.

References

Ajjan, R. A., and Grant, P. J. (2008) The cardiovascular safety of rosiglitazone. *Expert Opinion on Drug Safety*, 7(4), 367–376. DOI: 10.1517/14740338.7.4.367. 46

Aleman-Meza, B., Halaschek-Weiner, C., Arpinar, I. B., Ramakrishnan, C., and Sheth, A. P. (2005) Ranking complex relationships on the semantic web. *IEEE Internet Computing*, 9(3), 37–44. 14, 35

Andrews, R. C., Rooyackers, O., and Walker, B. R. (2003) Effects of the 11β-hydroxysteroid dehydrogenase inhibitor carbenoxolone on insulin sensitivity in men with type 2 diabetes. *Journal of Clinical Endocrinology and Metabolism*, 88(1), 285–291. DOI: 10.1210/jc.2002-021194. 32

Antezana, E., Blondé, W., Egaña, M., Rutherford, A., Stevens, R., De Baets, B., and Kuiper, M. (2009) BioGateway: A semantic systems biology tool for the life sciences. *BMC Bioinformatics*, 10(Suppl 10), S11. DOI: 10.1186/1471-2105-10-S10-S11. 20

Anyanwu, K., and Sheth, A. (2002) The ρ operator: Discovering and ranking associations on the semantic web. *ACM SIGMOD Record*, 31, 42–47. DOI: 10.1145/637411.637418. 72, 73

Anyanwu, K., Maduko, A., and Sheth, A. (2005) SemRank: Ranking complex relationship search results on the semantic web. In *Proceedings of the 14th international conference on World Wide Web* (pp. 117–127). ACM. DOI: 10.1145/1060745.1060766. 13

Ashburn, T. T., and Thor, K. B. (2004) Drug repositioning: Identifying and developing new uses for existing drugs. *Nature Reviews Drug discovery*, 3(8), 673–683. DOI: 10.1038/nrd1468. 91

Ashburner, M., Ball, C. A., Blake, J. A., Botstein, D., Butler, H., Cherry, J. M., and Sherlock, G. (2000) Gene Ontology: Tool for the unification of biology. *Nature Genetics*, 25(1), 25–29. DOI: 10.1038/75556. 10, 35

Baitaluk, M., and Ponomarenko, J. (2010) Semantic integration of data on transcriptional regulation. *Bioinformatics*, 26(13), 1651–1661. DOI: 10.1093/bioinformatics/btq231. 36

Barabási, A. L., Gulbahce, N., and Loscalzo, J. (2011) Network medicine: A network-based approach to human disease. *Nature Reviews Genetics*, 12(1), 56–68. DOI: 10.1038/nrg2918. 1

Bechhofer, S., Ainsworth, J. D., Bhagat, J., Buchan, I. E., Couch, P. A., Cruickshank, D., and Sufi, S. (2010) Why linked data is not enough for scientists. In *e-Science, IEEE Computer Society* (pp. 300–307). DOI: 10.1016/j.future.2011.08.004. 9

Belleau, F., Nolin, M. A., Tourigny, N., Rigault, P., and Morissette, J. (2008) Bio2RDF: Towards a mashup to build bioinformatics knowledge systems. *Journal of Biomedical Informatics*, *41*(5), 706–716. DOI: 10.1016/j.jbi.2008.03.004. 9, 20, 35

Bennet, A. M., Di Angelantonio, E., Ye, Z., Wensley, F., Dahlin, A., Ahlbom, A., and Danesh, J. (2007) Association of Apolipoprotein E Genotypes With Lipid Levels and Coronary Risk. *The Journal of the American Medical Association 298*, 1300–1311. DOI: 10.1001/jama.298.11.1300. 85

Benson, M. L., Smith, R. D., Khazanov, N. A., Dimcheff, B., Beaver, J., Dresslar, P., and Carlson, H. A. (2008) Binding MOAD, a high-quality protein–ligand database. *Nucleic Acids Research*, *36*(Database issue), D674-D678. DOI: 10.1093/nar/gkm911. 21, 24

Berners-Lee, T., Hendler, J., and Lassila, O. (2001) The Semantic Web. *Scientific American*, *284*(5), 34–43. DOI: 10.1038/scientificamerican0501-34. 3, 20

Bizer, C., and Cyganiak, R. (2006) D2R server-publishing relational databases on the semantic web. In *Proceeding of the 5th International Semantic Web Conference*. IEEE. 7, 21

Bleakley, K., and Yamanishi, Y. (2009) Supervised prediction of drug–target interactions using bipartite local models. *Bioinformatics*, *25*(18), 2397–2403. DOI: 10.1093/bioinformatics/btp433. 13, 92

Blei, D. M., and McAuliffe, J. D. (2010) Supervised Topic Models. *arXiv preprint arXiv:1003.0783.* 50

Blei, D. M., Ng, A. Y., and Jordan, M. I. (2003) Latent dirichlet allocation. *Journal of Machine Learning Research*, *3*, 993–1022. 50

Blei, D. M., Franks, K., Jordan, M. I., and Mian, I. S. (2006) Statistical modeling of biomedical corpora: Mining the Caenorhabditis Genetic Center Bibliography for genes related to life span. *BMC Bioinformatics*, *7*, 250. DOI: 10.1186/1471-2105-7-250. 50

Brinkman, R. R., Courtot, M., Derom, D., Fostel, J. M., He, Y., Lord, P., and Zheng, J. (2010) Modeling biomedical experimental processes with OBI. *J Biomed Semantics*, *1*(Suppl 1), S7. DOI: 10.1186/2041-1480-1-S1-S7. 10, 36

Bruford, E. A., Lush, M. J., Wright, M. W., Sneddon, T. P., Povey, S., and Birney, E. (2008) The HGNC Database in 2008: A resource for the human genome. *Nucleic Acids Research*, *36*(Database issue), D445-D448. DOI: 10.1093/nar/gkm881. 24

Butcher, E. C. (2005) Can cell systems biology rescue drug discovery?. *Nature Reviews Drug Discovery*, *4*, 461–467. DOI: 10.1038/nrd1754. 71

Campillos, M., Kuhn, M., Gavin, A. C., Jensen, L. J., and Bork, P. (2008) Drug target identification using side-effect similarity. *Science*, *321*(5886), 263–266. DOI: 10.1126/science.1158140. 91, 102

Chang, P. Y., Draheim, K., Kelliher, M. A., and Miyamoto, S. (2006) NFKB1 is a direct target of the TAL1 oncoprotein in human T leukemia cells. *Cancer Research*, *66*(12), 6008–6013. DOI: 10.1158/0008-5472.CAN-06-0194. 89

Chen, B., and Wild, D. J. (2010) PubChem BioAssays as a data source for predictive models. *Journal of Molecular Graphics and Modelling*, *28*(5), 420–426. DOI: 10.1016/j.jmgm.2009.10.001. 21

Chen, B., Ding, Y., Wang, H., Wild, D. J., Dong, X., Sun, Y., and Sankaranarayanan, M. (2010a) Chem2Bio2RDF: A Linked Open Data Portal for Systems Chemical Biology. In *Proceedings of 2010 IEEE/WIC/ACM International Conference on Web Intelligence and Intelligent Agent Technology* (pp. 232–239). IEEE. DOI: 10.1109/WI-IAT.2010.183. 35, 43

Chen, B., Dong, X., Jiao, D., Wang, H., Zhu, Q., Ding, Y., and Wild, D. J. (2010b) Chem2Bio2RDF: A semantic framework for linking and data mining chemogenomic and Systems Chemical Biology data. *BMC Bioinformatics*, *11*(1), 255. DOI: 10.1186/1471-2105-11-255. 10, 42, 43

Chen, B., Ding, Y., and Wild, D. J. (2012a) Improving integrative searching of systems chemical biology data using semantic annotation. *Journal of Cheminformatics*, *4*(1), 6. DOI: 10.1186/1758-2946-4-6. 14

Chen, B., Wild, D., and Guha, R. (2009a) PubChem as a source of polypharmacology. *Journal of Chemical Information and Modeling*, *49*(9), 2044–2055. DOI: 10.1021/ci9001876. 20, 21, 25

Chen, H., Ding, L., Wu, Z., Yu, T., Dhanapalan, L., and Chen, J. Y. (2009b) Semantic web for integrated network analysis in biomedicine. *Briefings in Bioinformatics*, *10*(2), 177–192. DOI: 10.1093/bib/bbp002. 13

Chen, J. Y., Mamidipalli, S., and Huan, T. (2009c) HAPPI: An online database of comprehensive human annotated and predicted protein interactions. *BMC Genomics*, *10*(Suppl 1), S16. DOI: 10.1186/1471-2164-10-S1-S16. 16

Chen, B., Ding, Y., and Wild, D. J. (2012b) Assessing drug target association using semantic linked data. *PLoS Computational Biology*, *8*(7), e1002574. DOI: 10.1371/journal.pcbi.1002574. 13, 14, 15

Chepelev, L. L., Riazanov, A., Kouznetsov, A., Low, H. S., Dumontier, M., and Baker, C. J. (2011) Prototype semantic infrastructure for automated small molecule classification and annotation in lipidomics. *BMC Bioinformatics*, *12*(1), 303. DOI: 10.1186/1471-2105-12-303. 14

Cheung, K. H., Yip, K. Y., Smith, A., Masiar, A., and Gerstein, M. (2005) YeastHub: A semantic web use case for integrating data in the life sciences domain. *Bioinformatics*, *21*(suppl 1), i85–i96. DOI: 10.1093/bioinformatics/bti1026. 20

Choi, J., Davis, M. J., Newman, A. F., and Ragan, M. A. (2010) A semantic web ontology for small molecules and their biological targets. *Journal of Chemical Information and Modeling*, *50*(5), 732–741. DOI: 10.1021/ci900461j. 10, 36, 37

Chu, L. H., and Chen, B. S. (2008) Construction of a cancer-perturbed protein-protein interaction network for discovery of apoptosis drug targets. *BMC Systems Biology*, *2*, 56. DOI: 10.1186/1752-0509-2-56. 62

Ciccarese, P., Wu, E., Wong, G., Ocana, M., Kinoshita, J., Ruttenberg, A., and Clark, T. (2008) The SWAN biomedical discourse ontology. *Journal of Biomedical Informatics*, *41*(5), 739–751. DOI: 10.1016/j.jbi.2008.04.010. 10, 36

Cohen, K. B., and Hunter, L. (2004) Natural language processing and systems biology. *Artificial Intelligence Methods and Tools for Systems Biology* (pp. 147–173). Springer Netherlands. DOI: 10.1007/978-1-4020-5811-0_9. 49

Cohen, J. S. (2006) Risks of troglitazone apparent before approval in USA. *Diabetologia*, *49*(6), 1454–1455. DOI: 10.1007/s00125-006-0245-0. 46

Coupet, J., Fisher, S. K., Rauh, C. E., Lai, F., and Beer, B. (1985) Interaction of amoxapine with muscarinic cholinergic receptors: An in vitro assessment. *European Journal of Pharmacology*, *112*(2), 231–235. DOI: 10.1016/0014-2999(85)90500-X. 31

Dani, C., Bertini, G., Pezzati, M., Poggi, C., Guerrini, P., Martano, C., and Rubaltelli, F. F. (2005) Prophylactic ibuprofen for the prevention of intraventricular hemorrhage among preterm infants: A multicenter, randomized study. *Pediatrics*, *115*(6), 1529–1535. DOI: 10.1542/peds.2004-1178. 86

Das, S., Sundara, S., and Cyganiak, R. (2012) R2RML: RDB to RDF mapping language. *W3C Recommendation*. Available at: http://www.w3.org/TR/r2rml/ 7

Davis, J., and Goadrich, M. (2006) The relationship between Precision-Recall and ROC curves. In *Proceedings of the 23rd international conference on Machine learning* (pp. 233–240). ACM. DOI: 10.1145/1143844.1143874. 97

Davis, A. P., King, B. L., Mockus, S., Murphy, C. G., Saraceni-Richards, C., Rosenstein, M., and Mattingly, C. J. (2011) The Comparative Toxicogenomics Database: update 2011. *Nucleic Acids Research*, *39*(Database issue), D1067-D1072. DOI: 10.1093/nar/gkq813. 20, 24

Dawson, M. A., Prinjha, R. K., Dittmann, A., Giotopoulos, G., Bantscheff, M., Chan, W. I., and Kouzarides, T. (2011) Inhibition of BET recruitment to chromatin as an effective treatment for MLL-fusion leukaemia. *Nature, 478*(7370), 529–533. DOI: 10.1038/nature10509. 88

de Matos, P., Alcántara, R., Dekker, A., Ennis, M., Hastings, J., Haug, K., and Steinbeck, C. (2010) Chemical entities of biological interest: An update. *Nucleic Acids Research, 38*(Database issue), D249-D254. DOI: 10.1093/nar/gkp886. 24

Degtyarenko, K., de Matos, P., Ennis, M., Hastings, J., Zbinden, M., McNaught, A., and Ashburner, M. (2008) ChEBI: A database and ontology for chemical entities of biological interest. *Nucleic Acids Research, 36*(Database issue), D344–50. DOI: 10.1093/nar/gkm791. 10, 36

Demir, E., Cary, M. P., Paley, S., Fukuda, K., Lemer, C., Vastrik, I., and Finney, A. (2010) The BioPAX community standard for pathway data sharing. *Nature Biotechnology, 28*(9), 935–942. DOI: 10.1038/nbt.1666. 10, 36, 37

Dentler, K., Cornet, R., ten Teije, A., and de Keizer, N. (2011) Comparison of reasoners for large ontologies in the OWL 2 EL profile. *Semantic Web Journal, 2*(2), 71–87. DOI: 10.3233/SW-2011-0034. 13

D'Eustachio, P. (2011) Reactome knowledgebase of human biological pathways and processes. *Bioinformatics for Comparative Proteomics, 694*,49–61. DOI: 10.1007/978-1-60761-977-2_4. 24

Dong, X., Ding, Y., Wang, H., Chen, B., and Wild, D. (2010) Chem2Bio2RDF Dashboard: Ranking semantic associations in Systems Chemical Biology space. In *Workshopp of Future of the Web in Collaboratice Science (FWCS).* 35

Dudley, J. T., Deshpande, T., and Butte, A. J. (2011) Exploiting drug-disease relationships for computational drug repositioning. *Briefings in Bioinformatics,12*(4), 303–311. DOI: 10.1093/bib/bbr013. 91

Dumontier, M., and Villanueva-Rosales, N. (2009) Towards pharmacogenomics knowledge discovery with the semantic web. *Briefings in Bioinformatics, 10*(2), 153–163. DOI: 10.1093/bib/bbn056. 10, 36

Durant, J. L., Leland, B. A., Henry, D. R., and Nourse, J. G. (2002) Reoptimization of MDL keys for use in drug discovery. *Journal of Chemical Information and Computer Sciences, 42*(6), 1273–1280. DOI: 10.1021/ci010132r. 88

Editorial. (2008) Networking chemical biology. *Nature Chemical Biology, 4*(11), 633. 1

Eilbeck, K., Lewis, S. E., Mungall, C. J., Yandell, M., Stein, L., Durbin, R., and Ashburner, M. (2005) The Sequence Ontology: A tool for the unification of genome annotations. *Genome Biology*, *6*(5), R44. DOI: 10.1186/gb-2005-6-5-r44. 10, 35

Ekins, S., Chang, C., Mani, S., Krasowski, M. D., Reschly, E. J., Iyer, M., and Bachmann, K. (2007) Human pregnane X receptor antagonists and agonists define molecular requirements for different binding sites. *Molecular Pharmacology*, *72*(3), 592–603. DOI: 10.1124/mol.107.038398. 45

Eronen, L., and Toivonen, H. (2012) Biomine: Predicting links between biological entities using network models of heterogeneous databases. *BMC Bioinformatics*, *13*(1), 119. DOI: 10.1186/1471-2105-13-119. 14, 15, 16

Ertel, W., Morrison, M. H., Meldrum, D. R., Ayala, A., and Chaudry, I. H. (1992) Ibuprofen restores cellular immunity and decreases susceptibility to sepsis following hemorrhage. *Journal of Surgical Research*, *53*(1), 55–61. DOI: 10.1016/0022-4804(92)90013-P. 86

Evans, W. E., and Johnson, J. A. (2001) Pharmacogenomics: The inherited basis for inter individual differences in drug response. *Annual Review of Genomics and Human Genetics*, *2*(1), 9–39. DOI: 10.1146/annurev.genom.2.1.9. 1

Fawcett, T. (2006) An introduction to ROC analysis. *Pattern Recognition Letters*, *27*(8), 861–874. DOI: 10.1016/j.patrec.2005.10.010. 97

Feldman, R., Regev, Y., Hurvitz, E., and Finkelstein-Landau, M. (2003) Mining the biomedical literature using semantic analysis and natural language processing techniques. *Biosilico*, *1*(2), 69–80. DOI: 10.1016/S1478-5382(03)02330-8. 49

Ferreira, J. D., and Couto, F. M. (2010) Semantic similarity for automatic classification of chemical compounds. *PLoS Computational Biology*, *6*(9), e1000937. DOI: 10.1371/journal.pcbi.1000937. 92

Fishman, M. C., and Porter, J. A. (2005) Pharmaceuticals: A new grammar for drug discovery. *Nature*, *437*(7058), 491–493. DOI: 10.1038/437491a. 1

Frazier, J. L., Pradilla, G., Wang, P. P., and Tamargo, R. J. (2004) Inhibition of cerebral vasospasm by intracranial delivery of ibuprofen from a controlled-release polymer in a rabbit model of subarachnoid hemorrhage. *Journal of Neurosurgery*, *101*(1), 93–98. DOI: 10.3171/jns.2004.101.1.0093. 86

Frijters, R., van Vugt, M., Smeets, R., van Schaik, R., de Vlieg, J., and Alkema, W. (2010) Literature mining for the discovery of hidden connections between drugs, genes and diseases. *PLoS Computational Biology*, *6*, e1000943. DOI: 10.1371/journal.pcbi.1000943. 72

Funk, C., Ponelle, C., Scheuermann, G., and Pantze, M. (2001) Cholestatic potential of trogli-tazone as a possible factor contributing to troglitazone-induced hepatotoxicity: In vivo and in vitro interaction at the canalicular bile salt export pump (Bsep) in the rat. *Molecular Pharmacology, 59*(3), 627–635. 44

Garcia-Serna, R., Ursu, O., Oprea, T. I., and Mestres, J. (2010) iPHACE: Integrative navigation in pharmacological space. *Bioinformatics, 26*(7), 985–986. DOI: 10.1093/bioinformatics/btq061. 91

Gaulton, A., Bellis, L. J., Bento, A. P., Chambers, J., Davies, M., Hersey, A., and Overington, J. P. (2012) ChEMBL: A large-scale bioactivity database for drug discovery. *Nucleic Acids Research, 40*(Database issue), D1100-D1107. DOI: 10.1093/nar/gkr777. 21, 24, 95

Goh, K. I., Cusick, M. E., Valle, D., Childs, B., Vidal, M., and Barabási, A. L. (2007) The human disease network. *Proceedings of the National Academy of Sciences, 104*(21), 8685–8690. DOI: 10.1073/pnas.0701361104. 24

Günther, S., Kuhn, M., Dunkel, M., Campillos, M., Senger, C., Petsalaki, E., and Preissner, R. (2008) SuperTarget and Matador: Resources for exploring drug-target relationships. *Nucleic Acids Research, 36*(Database issue), D919-D922. DOI: 10.1093/nar/gkm862. 21, 24

Haarslev, V., and Möller, R. (2001) RACER System Description. In *Proceedings of the First International Joint Conference on Automated Reasoning* (pp. 701–706). Springer-Verlag. DOI: 10.1007/3-540-45744-5_59. 12

Han, J., Sun, Y., Yan, X., and Yu, P. S. (2012) Mining knowledge from data: An information network analysis approach. In *Proceedings of the IEEE 28th International Conference on Data Engineering (ICDE)* (pp. 1214–1217). IEEE. 13

Harland, L. and Gaulton, A. (2009) Drug target central. *Expert Opinion on Drug Discovery, 4*(8), 857–872. DOI: 10.1517/17460440903049290. 40

He, B., Tang, J., Ding, Y., Wang, H., Sun, Y., Shin, J. H., and Wild, D. J. (2011) Mining relational paths in integrated biomedical data. *PLoS One, 6*(12), e27506. DOI: 10.1371/journal.pone.0027506. 85, 95

Heim, P., Lohmann, S., and Stegemann, T. (2010) Interactive relationship discovery via the semantic web. *Lecture Notes in Computer Science, 6088/2010*: 303–317 DOI: 10.1007/978-3-642-13486-9_21. 35

Hoehndorf, R., Dumontier, M., and Gkoutos, G. V. (2012) Identifying aberrant pathways through integrated analysis of knowledge in pharmacogenomics. *Bioinformatics, 28*(16), 2169–2175. DOI: 10.1093/bioinformatics/bts350. 14

Hofmann, T. (1999) Probabilistic latent semantic indexing. In *Proceedings of the 22nd annual international ACM SIGIR conference on Research and Development in Information Retrieval* (pp. 50–57). ACM. DOI: 10.1145/312624.312649. 50

Holford, M. E., Khurana, E., Cheung, K. H., and Gerstein, M. (2010) Using semantic web rules to reason on an ontology of pseudogenes. *Bioinformatics, 26*(12), i71-i78. DOI: 10.1093/bioinformatics/btq173. 36

Hopkins, A. L. (2008) Network pharmacology: The next paradigm in drug discovery. *Nature Chemical Biology, 4*(11), 682–690. DOI: 10.1038/nchembio.118. 1, 19

Horridge, M., and Bechhofer, S. (2008) The OWL API: A Java API for working with OWL 2 ontologies. In *Proceedings of the OWLED 2009 Workshop on OWL: Experiences and Directions. CEUR Workshop Proceedings* (Vol. 529). 13, 41

Hwang, T., Zhang, W., Xie, M., Liu, J., and Kuang, R. (2011) Inferring disease and gene set associations with rank coherence in networks. *Bioinformatics (Oxford, England), 27*(19), 2692–2699. DOI: 10.1093/bioinformatics/btr463. 72

Ivchenko, O., Younesi, E., Shahid, M., Wolf, A., Müller, B., and Hofmann-Apitius, M. (2011) PLIO: An ontology for formal description of protein-ligand interactions. *Bioinformatics, 27*(12), 1684–1690. DOI: 10.1093/bioinformatics/btr256. 10, 36

Jacob, L., and Vert, J. P. (2008) Protein-ligand interaction prediction: An improved chemogenomics approach. *Bioinformatics, 24*(19), 2149–2156. DOI: 10.1093/bioinformatics/btn409. 97

Jeh, G., and Widom, J. (2002) SimRank: A measure of structural-context similarity. In *Proceedings of the eighth ACM SIGKDD international conference on Knowledge Discovery and Data Mining* (pp. 538–543). ACM. DOI: 10.1145/775047.775126. 13

Jentzsch, A., Zhao, J., Hassanzadeh, O., Cheung, K. H., Samwald, M., and Andersson, B. (2009) Linking open drug data. In *Triplification Challenge of the International Conference on Semantic Systems* (pp. 3–6). 9, 20, 35

Jeong, H., Mason, S. P., Barabási, A. L., and Oltvai, Z. N. (2001) Lethality and centrality in protein networks. *Nature, 411*(6833), 41–42. DOI: 10.1038/35075138. 13

Jupp, S., Stevens, R., and Hoehndorf, R. (2012) Logical Gene Ontology Annotations (GOAL): Exploring gene ontology annotations with OWL. *Journal of Biomedical Semantics, 3*(Suppl 1), S3. 14

Kanehisa, M., Goto, S., Hattori, M., Aoki-Kinoshita, K. F., Itoh, M., Kawashima, S., and Hirakawa, M. (2006) From genomics to chemical genomics: New developments in KEGG. *Nucleic Acids Research, 34*(Database issue), D354-D357. DOI: 10.1093/nar/gkj102. 20, 24

Keiser, M. J., Roth, B. L., Armbruster, B. N., Ernsberger, P., Irwin, J. J., and Shoichet, B. K. (2007) Relating protein pharmacology by ligand chemistry. *Nature Biotechnology*, *25*(2), 197–206. DOI: 10.1038/nbt1284. 91

Keiser, M. J., Setola, V., Irwin, J. J., Laggner, C., Abbas, A. I., Hufeisen, S. J., and Roth, B. L. (2009) Predicting new molecular targets for known drugs. *Nature*, *462*(7270), 175–181. DOI: 10.1038/nature08506. 91, 97, 102

Keith, C. T., Borisy, A. A., and Stockwell, B. R. (2005) Multicomponent therapeutics for networked systems. *Nature Reviews Drug Discovery*, *4*(1), 71–78. DOI: 10.1038/nrd1609. 28

Khatri, P., Sirota, M., and Butte, A. J. (2012) Ten years of pathway analysis: Current approaches and outstanding challenges. *PLoS Computational Biology*, *8*(2), e1002375. DOI: 10.1371/journal.pcbi.1002375. 14, 16

Kinnings, S. L., Liu, N., Buchmeier, N., Tonge, P. J., Xie, L., and Bourne, P. E. (2009) Drug discovery using chemical systems biology: Repositioning the safe medicine Comtan to treat multidrug and extensively drug resistant tuberculosis. *PLoS Computational Biology*, *5*(7), e1000423. DOI: 10.1371/journal.pcbi.1000423. 91

Klein, T. E., Chang, J. T., Cho, M. K., Easton, K. L., Fergerson, R., Hewett, M., and Altman, R. B. (2001) Integrating genotype and phenotype information: An overview of the PharmGKB project. *Pharmacogenomics Journal*, *1*(3), 167–170. DOI: 10.1038/sj.tpj.6500035. 21, 24

Kola, I., and Landis, J. (2004) Can the pharmaceutical industry reduce attrition rates?. *Nature Reviews Drug Discovery*, *3*(8), 711–716. DOI: 10.1038/nrd1470. 1

Kuhn, M., Szklarczyk, D., Franceschini, A., Campillos, M., von Mering, C., Jensen, L. J., and Bork, P. (2010a) STITCH 2: An interaction network database for small molecules and proteins. *Nucleic Acids Research*, *38*(Database issue), D552-D556. DOI: 10.1093/nar/gkp937. 91

Kuhn, M., Campillos, M., Letunic, I., Jensen, L. J., and Bork, P. (2010) A side effect resource to capture phenotypic effects of drugs. *Molecular Systems Biology*, *6*, 343. DOI: 10.1038/msb.2009.98. 24

Kwan, E. P., Gao, X., Leung, Y. M., and Gaisano, H. Y. (2007) Activation of Exchange Protein Directly Activated by Cyclic Adenosine Monophosphate and Protein Kinase A Regulate Common and Distinct Steps in Promoting Plasma Membrane Exocytic and Granule-to-Granule Fusions in Rat Islet [beta] Cells. *Pancreas*, *35*(3), e45-e54. DOI: 10.1097/mpa.0b013e318073d1c9. 67

Lamb, J., Crawford, E. D., Peck, D., Modell, J. W., Blat, I. C., Wrobel, M. J., and Golub, T. R. (2006) The Connectivity Map: Using gene-expression signatures to connect small molecules, genes, and disease. *Science*, *313*(5795), 1929–1935. DOI: 10.1126/science.1132939. 75, 92, 103

Lessard, É., Yessine, M. A., Hamelin, B. A., O'Hara, G., LeBlanc, J., and Turgeon, J. (1999) Influence of CYP2D6 activity on the disposition and cardiovascular toxicity of the antidepressant agent venlafaxine in humans. *Pharmacogenetics and Genomics*, 9(4), 435–443. 47

Li, J., Zhu, X., and Chen, J. Y. (2008) Mining disease-specific molecular association profiles from biomedical literature: A case study. In *Proceedings of the ACM Symposium on Applied Computing* (pp. 1287–1291). ACM. DOI: 10.1145/1363686.1363984. 62

Li, Y. Y., An, J., and Jones, S. J. (2011) A computational approach to finding novel targets for existing drugs. *PLoS Computational Biology*, 7(9), e1002139. DOI: 10.1371/journal.pcbi.1002139. 91

Liben-Nowell, D., and Kleinberg, J. (2007) The link-prediction problem for social networks. *Journal of the American society for information science and technology*, 58(7), 1019–1031. DOI: 10.1002/asi.20591. 13, 99, 100

Liekens, A. M., De Knijf, J., Daelemans, W., Goethals, B., De Rijk, P., and Del-Favero, J. (2011) BioGraph: Unsupervised biomedical knowledge discovery via automated hypothesis generation. *Genome Biology*, 12(6), R57. DOI: 10.1186/gb-2011-12-6-r57. 14

Liu, T., Lin, Y., Wen, X., Jorissen, R. N., and Gilson, M. K. (2007) BindingDB: A web-accessible database of experimentally determined protein-ligand binding affinities. *Nucleic Acids Research*, 35(Database issue), D198-D201. DOI: 10.1093/nar/gkl999. 20, 24

Liu, Y., Hu, B., Fu, C., and Chen, X. (2010) DCDB: Drug combination database. *Bioinformatics*, 26(4), 587–588. DOI: 10.1093/bioinformatics/btp697. 24

Luciano, J. S., Andersson, B., Batchelor, C., Bodenreider, O., Clark, T., Denney, C. K., and Dumontier, M. (2011) The Translational Medicine Ontology and Knowledge Base: Driving personalized medicine by bridging the gap between bench and bedside. *Journal of Biomedical Semantics*, 2(Suppl 2), S1. DOI: 10.1186/2041-1480-2-S2-S1. 10, 36

Marshall, M. S., Boyce, R., Deus, H. F., Zhao, J., Willighagen, E. L., Samwald, M., and Stephens, S. (2012) Emerging practices for mapping and linking life sciences data using RDF—A case series. *Journal of Web Semantics*, 14, 2–13. DOI: 10.1016/j.websem.2012.02.003. 9

Mestres, J., Gregori-Puigjané, E., Valverde, S., and Sole, R. V. (2008) Data completeness—the Achilles heel of drug-target networks. *Nature Biotechnology*, 26(9), 983–984. DOI: 10.1038/nbt0908-983. 109

Mironov, V., Seethappan, N., Blondé, W., Antezana, E., Splendiani, A., and Kuiper, M. (2012) Gauging triple stores with actual biological data. *BMC Bioinformatics*, 13(Suppl 1), S3. DOI: 10.1186/1471-2105-13-S1-S3. 8

Mörchen, F., Dejori, M., Fradkin, D., Etienne, J., Wachmann, B., and Bundschus, M. (2008) Anticipating annotations and emerging trends in biomedical literature. In *Proceedings of the 14th ACM SIGKDD international conference on Knowledge Discovery and Data Mining* (pp. 954–962). ACM. DOI: 10.1145/1401890.1402004. 51

Muin, M., Fontelo, P., and Ackerman, M. (2006) PubMed Interact: an interactive search application for MEDLINE/PubMed. In *Proceedings of the AMIA Annual Symposium* (Vol. 2006, p. 1039). American Medical Informatics Association. 49

Myles, S., Hradetzky, E., Engelken, J., Lao, O., Nürnberg, P., Trent, R. J., and Stoneking, M. (2007) Identification of a candidate genetic variant for the high prevalence of type II diabetes in Polynesians. *European Journal of Human Genetics*, *15*, 584–589. DOI: 10.1038/sj.ejhg.5201793. 67

Natale, D. A., Arighi, C. N., Barker, W. C., Blake, J., Chang, T. C., Hu, Z., and Wu, C. H. (2007) Framework for a protein ontology. *BMC Bioinformatics*, *8*(Suppl 9), S1. DOI: 10.1186/1471-2105-8-S9-S1. 10, 35

Neumann, E. (2005) A life science semantic web: Are we there yet?. *Science STKE*, *2005*(283), pe22. DOI: 10.1126/stke.2832005pe22. 3, 20

Neumann, E. K., and Quan, D. (2006) BioDash: A Semantic Web dashboard for drug development. In *Pacific Symposium on Biocomputing. Pacific Symposium on Biocomputing* (pp. 176–187). 20

Nidhi, G. M., Davies, J. W., and Jenkins, J. L. (2006) Prediction of biological targets for compounds using multiple-category Bayesian models trained on chemogenomics databases. *Journal of Chemical Information and Modeling*, *46*(3), 1124–1133. DOI: 10.1021/ci060003g. 91

Nissen, S. E., and Wolski, K. (2007) Effect of rosiglitazone on the risk of myocardial infarction and death from cardiovascular causes. *New England Journal of Medicine*, *356*, 2457–2471. DOI: 10.1056/NEJMoa072761. 85

Noy, N. F., Shah, N. H., Whetzel, P. L., Dai, B., Dorf, M., Griffith, N., and Musen, M. A. (2009) BioPortal: Ontologies and integrated data resources at the click of a mouse. *Nucleic Acids Research*, *37*(suppl 2), W170-W173. DOI: 10.1093/nar/gkp440. 10, 36

O'Connor, K. A., and Roth, B. L. (2005) Finding new tricks for old drugs: an efficient route for public-sector drug discovery. *Nature Reviews Drug Discovery*, *4*(12), 1005–1014. DOI: 10.1038/nrd1900. 91

Oprea, T. I., Tropsha, A., Faulon, J. L., and Rintoul, M. D. (2007) Systems chemical biology. *Nature Chemical Biology*, *3*(8), 447–450. DOI: 10.1038/nchembio0807-447. 1, 19

Oprea, T. I., Nielsen, S. K., Ursu, O., Yang, J. J., Taboureau, O., Mathias, S. L., and Bologa, C. G. (2011) Associating drugs, targets and clinical outcomes into an integrated network affords a new platform for computer-aided drug repurposing. *Molecular Informatics*, *30*(2–3), 100-111. DOI: 10.1002/minf.201100023. 91

Orchard, S., Al-Lazikani, B., Bryant, S., Clark, D., Calder, E., Dix, I., and Thornton, J. (2011) Minimum information about a bioactive entity (MIABE). *Nature Reviews Drug Discovery*, *10*(9), 661–669. DOI: 10.1038/nrd3503. 10, 36

Page, L., Brin, S., Motwani, R., and Winograd, T. (1999) The PageRank citation ranking: Bringing order to the web. Technical Report 1999–66, Stanford University. 13

Pammolli, F., Magazzini, L., and Riccaboni, M. (2011) The productivity crisis in pharmaceutical RandD. *Nature Reviews Drug Discovery*, *10*(6), 428–438. DOI: 10.1038/nrd3405. 1

Pentel, P. R., Asinger, R. W., and Benowitz, N. L. (1985) Propranolol antagonism of phenylpropanolamine-induced hypertension. *Clinical Pharmacology and Therapeutics*, *37*(5), 488–494. DOI: 10.1038/clpt.1985.77. 105

Perlman, L., Gottlieb, A., Atias, N., Ruppin, E., and Sharan, R. (2011) Combining drug and gene similarity measures for drug-target elucidation. *Journal of Computational Biology*, *18*(2), 133–145. DOI: 10.1089/cmb.2010.0213. 92, 97

Pradilla, G., Thai, Q. A., Legnani, F. G., Clatterbuck, R. E., Gailloud, P., Murphy, K. P., and Tamargo, R. J. (2005) Local delivery of ibuprofen via controlled-release polymers prevents angiographic vasospasm in a monkey model of subarachnoid hemorrhage. *Neurosurgery*, *57*(1), 184–190. DOI: 10.1227/01.NEU.0000163604.52273.28. 86

Prasad, T. K., Goel, R., Kandasamy, K., Keerthikumar, S., Kumar, S., Mathivanan, S., and Pandey, A. (2009) Human protein reference database—2009 update. *Nucleic Acids Research*, *37*(Database issue), D767-D772. DOI: 10.1093/nar/gkn892. 24

Qi, D., Ross, K., Andrew, H., Richard, B., and Larisa, S. (2010) An ontology for description of drug discovery investigations. *Journal of Integrative Bioinformatics*, *7*(3). DOI: 10.2390/biecoll-jib-2010-126. 10, 36

Qu, X. A., Gudivada, R. C., Jegga, A. G., Neumann, E. K., and Aronow, B. J. (2009) Inferring novel disease indications for known drugs by semantically linking drug action and disease mechanism relationships. *BMC Bioinformatics*, *10*(Suppl 5), S4. DOI: 10.1186/1471-2105-10-S5-S4. 10, 36

Quilitz, B., and Leser, U. (2008) Querying distributed RDF data sources with SPARQL. *The Semantic Web: Research and Applications* (pp. 524–538). Springer Berlin Heidelberg. DOI: 10.1007/978-3-540-68234-9_39. 8

Reuter, S., Charlet, J., Juncker, T., Teiten, M. H., Dicato, M., and Diederich, M. (2009) Effect of curcumin on nuclear factor κB signaling pathways in human chronic myelogenous K562 leukemia cells. *Annals of the New York Academy of Sciences, 1171*(1), 436–447. DOI: 10.1111/j.1749-6632.2009.04731.x. 89

Rhee, S. Y., Wood, V., Dolinski, K., and Draghici, S. (2008) Use and misuse of the gene ontology annotations. *Nature Reviews Genetics, 9*(7), 509–515. DOI: 10.1038/nrg2363. 94

Rosen-Zvi, M., Griffiths, T., Steyvers, M., and Smyth, P. (2004) The author-topic model for authors and documents. In *Proceedings of the 20th conference on Uncertainty in Artificial Intelligence* (pp. 487–494). AUAI Press. 50

Ruebenacker, O., Moraru, I. I., Schaff, J. C., and Blinov, M. L. (2007) Kinetic modeling using BioPAX ontology. In *Proceedigns of the IEEE International Conference on Bioinformatics and Biomedicine (BIBM 2007)* (pp. 339–348). IEEE. DOI: 10.1109/BIBM.2007.55. 37

Salwinski, L., Miller, C. S., Smith, A. J., Pettit, F. K., Bowie, J. U., and Eisenberg, D. (2004) The database of interacting proteins: 2004 update. *Nucleic Acids Research, 32*(Database issue), D449-D451. DOI: 10.1093/nar/gkh086. 24

Scheiber, J., Chen, B., Milik, M., Sukuru, S. C. K., Bender, A., Mikhailov, D., and Jenkins, J. L. (2009) Gaining insight into off-target mediated effects of drug candidates with a comprehensive systems chemical biology analysis. *Journal of Chemical Information and Modeling, 49*(2), 308–317. DOI: 10.1021/ci800344p. 19, 91

Schierz, A. C. (2009) Virtual screening of bioassay data. *Journal of Cheminformatics, 1*, 21. DOI: 10.1186/1758-2946-1-21. 16

Schriml, L.M., Arze, C., Nadendla, S., Chang, Y. W., Mazaitis, M., Felix, V., and Warren A. K. (2012) Disease ontology: A backbone for disease semantic integration. Nucleic Acids Research, 40(D1), D940-D946. DOI: 10.1093/nar/gkr972. 10, 36, 41

Shadbolt, N., Hall, W., and Berners-Lee, T. (2006) The semantic web revisited. *IEEE Intelligent Systems, 21*(3), 96–101. DOI: 10.1109/MIS.2006.62. 91

Shannon, P., Markiel, A., Ozier, O., Baliga, N. S., Wang, J. T., Ramage, D., and Ideker, T. (2003) Cytoscape: A software environment for integrated models of biomolecular interaction networks. *Genome Research, 13*(11), 2498–2504. DOI: 10.1101/gr.1239303. 95

Shearer, R., Motik, B., and Horrocks, I. (2008) HermiT: A Highly-Efficient OWL Reasoner. In *Proceedings of the CEUR Workshop of OWLED* (Vol. 432). 12

Sirin, E., Parsia, B., Grau, B. C., Kalyanpur, A., and Katz, Y. (2007) Pellet: A practical owl-dl reasoner. *Journal of Web Semantics, 5*(2), 51–53. DOI: 10.1016/j.websem.2007.03.004. 12, 41

Slater, T., Bouton, C., and Huang, E. S. (2008) Beyond data integration. *Drug Discovery Today*, *13*(13–14), 584-589. DOI: 10.1016/j.drudis.2008.01.008. 19

Smith, B., Ceusters, W., Klagges, B., Köhler, J., Kumar, A., Lomax, J., and Rosse, C. (2005) Relations in biomedical ontologies. *Genome Biology*, *6*(5), R46. DOI: 10.1186/gb-2005-6-5-r46. 10, 37

Smith, B., Ashburner, M., Rosse, C., Bard, J., Bug, W., Ceusters, W., and Lewis, S. (2007) The OBO Foundry: Coordinated evolution of ontologies to support biomedical data integration. *Nature Biotechnology*, *25*(11), 1251–1255. DOI: 10.1038/nbt1346. 10, 36

Sorger, P. K., Allerheiligen, S. R., Abernethy, D. R., Altman, R. B., Brouwer, K. L., Califano, A., and Ward, R. (2011) Quantitative and systems pharmacology in the post-genomic era: New approaches to discovering drugs and understanding therapeutic mechanisms. In *An NIH white paper by the QSP workshop group* (pp. 1–48). Bethesda: NIH. 1

Sun, Y., Han, J., Yan, X., Yu, P. S., and Wu, T. (2011) PathSim: Meta path-based top-K similarity search in heterogeneous information networks. *Proceedings of the VLDB Endowment*, *4*(11), 992–1003. 13, 14

Taboureau, O., Nielsen, S. K., Audouze, K., Weinhold, N., Edsgärd, D., Roque, F. S., and Oprea, T. I. (2011) ChemProt: A disease chemical biology database. *Nucleic Acids Research*, *39*(Database issue), D367-D372. DOI: 10.1093/nar/gkq906. 91

Tang, J., Zhang, J., Yao, L., Li, J., Zhang, L., and Su, Z. (2008) Arnetminer: extraction and mining of academic social networks. In *Proceedings of the 14th ACM SIGKDD international conference on Knowledge Discovery and Data Mining* (pp. 990–998). ACM. DOI: 10.1145/1401890.1402008. 50

Tanimoto, T. T. (1957) Tamimoto coefficient. *IBM Internal Report 17th*. 88

Tsarkov, D., and Horrocks, I. (2006) FaCT++ description logic reasoner: System description. In *Proceedings of the Third International Joint Conference on Automated Reasoning* (Vol. 4130, p. 292–297). Springer-Verlag. DOI: 10.1007/11814771_26. 12

Velankar, S., Alhroub, Y., Best, C., Caboche, M., Conroy, M. J., Dana, J. M., and Kleywegt, G. J. (2012) PDBe: Protein data bank in Europe. *Nucleic Acids Research*, *40*(Database issue), D445-D452. DOI: 10.1093/nar/gkr998. 24

Vidal, D., and Mestres, J. (2010) In silico receptorome screening of antipsychotic drugs. *Molecular Informatics*, *29*(6–7), 543-551. DOI: 10.1002/minf.201000055. 102

Visser, U., Abeyruwan, S., Vempati, U., Smith, R. P., Lemmon, V., and Schürer, S. C. (2011) BioAssay Ontology (BAO): A semantic description of bioassays and high-throughput screening results. *BMC Bioinformatics*, *12*(1), 257. DOI: 10.1186/1471-2105-12-257. 10, 36

Wale, N. (2011) Machine learning in drug discovery and development. *Drug Development Research*, *72*(1), 112–119. DOI: 10.1002/ddr.20407. 88

Wang, X., and McCallum, A. (2006) Topics over time: a non-Markov continuous-time model of topical trends. In *Proceedings of the 12th ACM SIGKDD International Conference on Knowledge Discovery and Data Mining* (pp. 424–433). ACM. DOI: 10.1145/1150402.1150450. 50

Wang, H., Klinginsmith, J., Dong, X., Lee, A. C., Guha, R., Wu, Y., and Wild, D. J. (2007a) Chemical data mining of the NCI human tumor cell line database. *Journal of Chemical Information and Modeling*, *47*(6), 2063–2076. DOI: 10.1021/ci700141x. 25

Wang, J., Zhou, J. Y., and Wu, G. S. (2007b) ERK-Dependent MKP-1–Mediated Cisplatin Resistance in Human Ovarian Cancer Cells. *Cancer Research*, *67*(24), 11933–11941. DOI: 10.1158/0008-5472.CAN-07-5185. 29

Wang, Y., Xiao, J., Suzek, T. O., Zhang, J., Wang, J., and Bryant, S. H. (2009) PubChem: a public information system for analyzing bioactivities of small molecules. *Nucleic Acids Research*, *37*(Web Server issue), W623. DOI: 10.1093/nar/gkp456. 20, 24

Wang, H., Ding, Y., Tang, J., Dong, X., He, B., Qiu, J., and Wild, D. J. (2011) Finding complex biological relationships in recent PubMed articles using Bio-LDA. *PLoS One*, *6*(3), e17243. DOI: 10.1371/journal.pone.0017243. 95

Wild, D. J. (2009) Mining large heterogeneous data sets in drug discovery. *Expert Opinion on Drug Discovery*, *4*, 995–1004. DOI: 10.1517/17460440903233738. 19

Wild, D. J., Ding, Y., Sheth, A. P., Harland, L., Gifford, E. M., and Lajiness, M. S. (2012) Systems Chemical Biology and the Semantic Web: What they mean for the future of drug discovery research. *Drug Discovery Yoday*, *17*(9), 469–474. DOI: 10.1016/j.drudis.2011.12.019. 3

Williams, A. J., Ekins, S., and Tkachenko, V. (2012a) Towards a gold standard: Regarding quality in public domain chemistry databases and approaches to improving the situation. *Drug Discovery Today*, *17*(13–14), 685–701. DOI: 10.1016/j.drudis.2012.02.013. 16

Williams, A. J., Harland, L., Groth, P., Pettifer, S., Chichester, C., Willighagen, E. L., and Mons, B. (2012b) Open PHACTS: Semantic interoperability for drug discovery. *Drug Discovery Today*, *17*(21), 1188–1198. DOI: 10.1016/j.drudis.2012.05.016. 10

Wishart, D. S., Knox, C., Guo, A. C., Shrivastava, S., Hassanali, M., Stothard, P., and Woolsey, J. (2006) DrugBank: A comprehensive resource for in silico drug discovery and exploration. *Nucleic Acids Research*, *34*(Database issue), D668-D672. DOI: 10.1093/nar/gkj067. 21, 24

Wolstencroft, K., Lord, P., Tabernero, L., Brass, A., and Stevens, R. (2006) Protein classification using ontology classification. *Bioinformatics*, *22*(14), e530-e538. DOI: 10.1093/bioinformatics/btl208. 14

Xie, L., Wang, J., and Bourne, P. E. (2007) In silico elucidation of the molecular mechanism defining the adverse effect of selective estrogen receptor modulators. *PLoS Computational Biology*, *3*(11), e217. DOI: 10.1371/journal.pcbi.0030217. 91

Xie, L., Li, J., Xie, L., and Bourne, P. E. (2009) Drug discovery using chemical systems biology: Identification of the protein-ligand binding network to explain the side effects of CETP inhibitors. *PLoS Computational Biology*, *5*(5), e1000387. DOI: 10.1371/journal.pcbi.1000387. 46, 91

Yamanishi, Y., Kotera, M., Kanehisa, M., and Goto, S. (2010) Drug-target interaction prediction from chemical, genomic and pharmacological data in an integrated framework. *Bioinformatics*, *26*(12), i246-i254. DOI: 10.1093/bioinformatics/btq176. 72, 97, 101

Yang, L., Wang, K., Chen, J., Jegga, A. G., Luo, H., Shi, L., and He, L. (2011) Exploring off-targets and off-systems for adverse drug reactions via chemical-protein interactome—clozapine-induced agranulocytosis as a case study. *PLoS Computational Biology*, *7*(3), e1002016. DOI: 10.1371/journal.pcbi.1002016. 91

Yıldırım, M. A., Goh, K. I., Cusick, M. E., Barabási, A. L., and Vidal, M. (2007) Drug-target network. *Nature Biotechnology*, *25*(10), 1119–1126. DOI: 10.1038/nbt1338. 13, 97

Young, D. W., Bender, A., Hoyt, J., McWhinnie, E., Chirn, G. W., Tao, C. Y., and Feng, Y. (2008) Integrating high-content screening and ligand-target prediction to identify mechanism of action. *Nature Chemical Biology*, *4*(1), 59–68. DOI: 10.1038/nchembio.2007.53. 1

Zhao, S., and Li, S. (2010) Network-based relating pharmacological and genomic spaces for drug target identification. *PLoS One*, *5*(7), e11764. DOI: 10.1371/journal.pone.0011764. 92, 97

Zhao, J., Miles, A., Klyne, G., and Shotton, D. (2009) Linked data and provenance in biological data webs. *Briefings in Bioinformatics*, *10*(2), 139–152. DOI: 10.1093/bib/bbn044. 7

Zheng, B., McLean, D. C., and Lu, X. (2006) Identifying biological concepts from a protein-related corpus with a probabilistic topic model. *BMC Bioinformatics*, *7*, 58. DOI: 10.1186/1471-2105-7-135. 50

Zhu, F., Han, B., Kumar, P., Liu, X., Ma, X., Wei, X., and Chen, Y. (2010) Update of TTD: therapeutic target database. *Nucleic Acids Research*, *38*(Database issue), D787-D791. DOI: 10.1093/nar/gkp1014. 21, 24

Zhu, Q., Sun, Y., Challa, S., Ding, Y., Lajiness, M. S., and Wild, D. J. (2011) Semantic inference using chemogenomics data for drug discovery. *BMC Bioinformatics*, *12*(1), 256. DOI: 10.1186/1471-2105-12-256. 12

Authors' Biographies

BIN CHEN

Bin Chen is currently a postdoctoral scholar at Stanford University. His primary interest is to develop tools and algorithms to identify new therapeutics from publicly available data sources. He received his Ph.D. in informatics at Indiana University, Bloomington. During his graduate school, he primarily developed three systems (i.e., Chem2Bio2RDF, Chem2Bio2OWL and SLAP) used to represent, integrate and mine semantic data for drug discovery. He interned in three pharmaceutical companies (i.e., Novartis, Pfizer and Merck) for two years during this graduate school. He has published over 20 scientific papers.

HUIJUN WANG

Huijun Wang is an Associate Principle Scientist in the Cheminformatics Department at Merck. She leads the competitive intelligence data integration and information retrieval project and also focuses on text mining and data mining of chemical and biological information related to drug discovery. She has worked in the pharmaceutical industry for many years, including four years at Pfizer before joining Merck. She managed the large-scale internal and external data integration, retrieval and mining effects. She graduated from Indiana University with a Ph.D. in Informatics and an M.S. in Computer Science and Cheminformatics.

YING DING

Dr. Ying Ding is an Associate Professor at School of Informatics and Computing, Indiana University. Previously, she worked as a senior researcher at the University of Innsbruck, Austria and as a researcher at the Free University of Amsterdam, the Netherlands and has been involved in various NIH and European-Union funded Semantic Web projects and has published 170+ papers in journals, conferences, and workshops. She is the co-editor of book series *Semantic Web Synthesis* by Morgan & Claypool Publisher. She is co-author of the book *Intelligent Information Integration in B2B Electronic Commerce,* published by Kluwer Academic Publishers, and co-author of book chapters in the book *Spinning the Semantic Web,* published by MIT Press, and *Towards the Semantic Web: Ontology-driven Knowledge Management,* published by Wiley. Dr. Ding is on the editorial board of four ISI indexed top journals in Information Science and Semantic Web. Her current research interest areas include social network analysis, Semantic Web, citation analysis, knowledge management, and application of Web Technology.

DAVID WILD

David Wild is an Associate Professor at Indiana University's School of Informatics and Computing (SOIC) where he is a Graduate Program Director for the new Data Science Program and leads the Cheminformatics and Chemogenomics Research Group (CCRG). He is the founder and Chief Executive Officer at the data science technology company Data2Discovery Inc. and consults through Wild Consulting and Innovations. He has around 100 research publications, edits the *Journal of Cheminformatics,* and maintains a variety of educational resources including LearnCheminformatics.com, SurvivingDisasters.net and the All Hazards Blog. He is an Emergency Medical Technician (EMT) and volunteer with Bloomington Township Fire Department and Argus Canine Search and Rescue.

Printed in the United States
by Baker & Taylor Publisher Services